ANTHROPOLOGICAL USE OF PSYCHOACTIVE SUBSTANCES THROUGHOUT HUMAN HISTORY

Cultivation and Safe Use of Magic Mushrooms

by

Elia Friedenthal

Copyright © 2020 Elia Friedenthal

All rights reserved.

Copyright © 2020 Elia Friedenthal

All rights reserved.

This report is towards furnishing precise and solid data concerning the point and issue secured. The production is sold with the possibility that the distributor isn't required to render bookkeeping, formally allowed, or something else, qualified administrations. On the off chance that exhortation is important, lawful, or proficient, a rehearsed individual in the calling ought to be requested.

The Statement of Principles which was also accepted by the Americans Bar Association Committee and the Publishers and Associations Committee and endorsed by the Board.

Not the slightest bit is it lawful to replicate, copy, or transmit any piece of this report in either electronic methods or the printed group. Recording of this distribution is carefully disallowed, and any capacity of this report is not permitted except if with composed authorization from the distributor. All rights held.

The data gave in this is expressed, to be honest, and predictable, in that any risk, as far as absentmindedness or something else, by any utilization or maltreatment of any approaches, procedures, or bearings contained inside is the singular and articulate obligation of the beneficiary peruser. By no means will any lawful obligation or fault be held against the distributor for any reparation, harms, or money related misfortune because of the data in this, either straightforwardly or by implication.

Particular creators claim all copyrights not held by the distributor.

The data in this is offered for educational purposes exclusively and is all-inclusive as so. The introduction of the data is without a contract or any sort of assurance confirmation.

The trademarks that are utilized are with no assent, and the distribution of the trademark is without consent or support by the trademark proprietor. All trademarks and brands inside this book are for explaining purposes just and are simply possessed by the proprietors, not partnered with this record.

TABLE OF CONTENT

INTRODUCTION 7

CULTURE AND SUBSTANCE ABUSE: IMPACT OF CULTURE AFFECTS TREATMENT 11

PSYCHOACTIVE SUBSTANCE 20

PSYCHOACTIVE SUBSTANCES: HARM REDUCTION POSITION STATEMENT 49

9 THINGS THAT MATTER ABOUT PSYCHOACTIVE DRUGS 52

RECREATIONAL USAGE OF PSYCHOACTIVE SUBSTANCE 61

Which are one of the most common entertainment drugs used? 66

USAGE OF PSYCHOACTIVE SUBSTANCE IN RELIGION OVER THE YEARS 81

SURPRISING MEDICAL USES FOR ILLICIT DRUGS 89

DEVISING MORE POTENT COMPOUNDS 96

UNDERSTANDING PSYCHOACTIVE SUBSTANCE ABUSE AND VIOLENCE 104

SUBSTANCE ABUSE TREATMENT OVER THE YEARS 123

THE TRIUMVIRATE REASONS WHY TEENS SEEK DRUG ABUSE 139

THE MIND AS WELL AS BODY RELATIONSHIP MENTAL ILLNESS WITH A METAPHYSICALLY EXPLANATION 141

CONCLUSION 145

INTRODUCTION

Human addiction to psychotropic and also the state of mind-changing substances is a recurring worldwide problem. While the phenomenon of psychedelic drug use has actually gotten considerable academic interest, there needs to be more comparative techniques right into this area. Evolutionary as well as anthropological techniques for comprehending human demand for psychotropic and also the state of mind changing materials use the vicwcr's various informative angles for evaluating this sensation. While transformative techniques investigate the transformative actions and also mechanisms of human ancestral psychotropic usage, anthropological methods highlight the religious and also social definitions of specific drug use. While each method extends different concepts, a relative method, which is used in this research study, may offer a novel as well as invaluable insight for understanding human demand for psychotropic as well as mood-altering compounds.

Our very early forefathers lived as hunter-gatherers and also- as shown by the society of human teams that maintained this lifestyle (e.g., Australian natives, Amazon Indians, or Kalahari desert Bushmen) - they unquestionably accumulated significant info on medicinal plants. Ötzi, the man whose icy body was recuperated in the Alps in 1991, lived concerning 3300 years BC, and also brought in his pouch a traveling drug store consisting of polypore fungi with hemostatic and anti-bacterial residential properties. After embracing a pastoral lifestyle, people might have observed the effects of psychedelic plants on their flocks. Custom has it that Ethiopian priests began steaming and also roasting coffee beans to stay awake with evenings of the petition after a shepherd saw exactly how his goats were frolicking after eating coffee hedges.

Hallucinogens and relevant materials become part of a class of medications called psychoactive drugs. These are so-called because they act upon the mind of the individual utilizing the drug, which may influence in differing levels the understandings, emotions, thoughts, along with the level of awareness of the individual.

The use of psychoactive materials dating to at least 10,000 years and the historical-cultural usage over the last 5,000 years have been archaeologically recorded. Some say that marketing, preparation, and the stresses on contemporary lives are some of the reasons why people use many psychoactive substances in their daily lives.

Biology recommends a transformative link between psychedelic plants and also pets as to why these chemicals, as well as their receptors, exist within the nervous system. During the 20th century, many federal governments throughout the world originally reacted to the usage of leisure drugs by banning them as well as making their use, supply, or trade a criminal infraction. Several federal governments have wrapped up that illegal medication use cannot be completely quit through criminalization. In some countries, there has actually been a step toward injury reduction by wellness services, where the use of immoral medicines is neither pardoned nor advertised, but services and also assistance are offered to make certain users have the unfavorable results of their immoral drug use reduced.

Schematically, psychedelic compounds have been used

(1) In spiritual ceremonies by clergymen;

(ii) For medicinal purposes; or

(iii) Enormously, as staple products, by huge sections of the population in a socially authorized way.

Alternatively, poppy (opium) and hemp (marijuana) originated in Eurasia.1 In comparison, alcohol can quickly be created by the activity of yeast on a variety of plants consisting of starch or sugar, as well as has actually been made use of by basically all cultures. The abrupt destructive impact of alcohol on North American indigenous societies might be explained by the truth that traditional patterns of usage had not been developed; an additional possible variable might be the lack of previous genetic choice operating on prone topics over centuries.

CULTURE AND SUBSTANCE ABUSE: IMPACT OF CULTURE AFFECTS TREATMENT

There have been countless meanings of society. Dwight Heath1 supplies a simple interpretation: "It [society] is a system of patterns of belief and also actions that shape the worldview of the participant of culture. Therefore, it acts as a guide for activity, a cognitive map, and a grammar for habits."

Substance abuse describes the misuse of alcohol and other medications, mostly immoral drugs; however, what is thought about "illicit" is typically culturally determined and can vary between social groups. The majority of cultural groups have used and abused alcohol and other drugs over the years, as well as codes of conduct in their alcohol and drug policies.

This book starts with a short overview of the historical history as well as context for the use and misuse of substances. An evaluation of the effect of society on the initiation, use, as well as abuse of substances complies with.

HISTORIC ROOTS, FRUITS, AND PLANTS

Alcohol and several other drugs have been utilized

for countless years. Alcoholic beverages have actually been fermented from a range of plants and also fruits, considering that a minimum of 4000 bc. Both red wine and beer were first made concerning the very same time in what is now Iraq and Iran. A few of the earliest referrals to using alcohol are discovered in old Sumerian clay tablets that contain recipes for making use of red wine as a solvent for medicines. Alcohol consumption in North America is little listened to prior to white arrival. In Mexico, the Aztec, the Pima / Papago, in the southwest of the USA, and the Allies, in the north as far as Alaska, have been reporting alcohol consumption alone. In the early American times, alcohols were added in larger quantities.

The Sumerians grew opium cobbles at the same time that certain groups distilled liquors, and named them the Hul Gil plant. Opium cobbles were used for the therapy of the bowel and pain in their medical buildings and, due to their psychological residential and commercial properties, also for the treatment of sedation and euphoria.

The very early Chinese immigration and the subsequent introduction of the heroin to teams of the urban minorities such as blacks and Hispanics have been identifiable in the US as a source of opium.

Cannabis is believed to have its beginnings more than 4000 years ago in China, as well as later in India. Before its psychedelic use, marijuana (hemp) was utilized as fiber, and traces of its use for fabric go back greater than 10,000 years in China.5 Hemp was expanded by George Washington at Mount Vernon and also was the 2nd biggest crop-- after cotton-- expanded in the South before the US Civil War. After World War I, Mexican laborers introduced Americans to smoking cannabis for its psychedelic residential or commercial properties.

More than 1500 compounds have been found in most psychoactive plants worldwide in the Americas,6 hallucinogens, cocaine, and cigarettes7 have become one of the major drugs used in the United States for 30 years. The cocaine was made in the Andes of South America. Tobacco in the New World has been used approximately 5000 bc. Once he landed in the Caribbean, Christopher Columbus noticed that he used tobacco to cure many diseases. In the next 150 years, Tobacco will spread rapidly all over the world. Peyote was used for religious events in northern Mexico, accompanied by the colonization of Native American tribes.

THE ROLE OF CULTURE IN SUBSTANCE USE AND ABUSE

An instance is the usage of alcohol by the old Aztecs prior to any type of call with white inhabitants. Their use of alcohol was greatly managed and also was only for ritualistic objectives. Peyote was utilized in a ceremonial set up to treat persistent alcohol dependency.

Initiation into too much compound use may happen during durations of quick social change, frequently among cultural teams that have had little direct exposure to medication and also have not created protective normative habits. Anomie, or loss of a healthy ethnic or cultural identity, may occur amongst native populations whose societies have been ruined by the substantial as well as a sudden influx of outside impact.

As a result of its low accessibility, few North American Indians had any kind of direct exposure to alcohol before the arrival of whites. On the western frontier, potent distilled alcoholic beverages ended up being extensively available, and also the only version Native Americans had was the intoxicated presence of the frontiersman.

Acculturation, the degree to which a specific identifies with his/her indigenous culture, is believed to be related to material use as well as misuse. Indigenous American senior citizens believe that numerous drug abuse issues relate to the loss of conventional society. Greater prices important use have actually been found face to faces that very closely identify with non-Native American worths, as well as the lowest prices, are discovered in bicultural people that are comfortable with both sets of cultural worths.

Immigrants leave the safety atmosphere of their

family members behind and are faced with a brand-new collection of cultural standards as well as values. These women might often assume the drinking behavior of the leading society, and, as an outcome, they boost their use of alcohol.

Accelerated alcohol abuse and illegal drug use in acculturated Hispanics is shown by recent research by the State of Washington 15 In the previous month, illegal and also decreased drug use was reported to be 7.2 percent compared with slightly less than 1 percent of unacculturated Hispanics and 6.4 percent of whites. Hispanics with no acculturation (recent immigrants) were more household-oriented and with lower prices for drug and alcohol use — clearly, the protective effects of the aboriginal cultural values.

ANALYSIS AND SCREENING

With the populace of the United States coming to be increasingly diverse, it is important to think about a person's cultural history when analyzing for drug abuse or reliance. The publication of DSM-IV was a vital transition in the application of cultural psychiatry principles due to the fact that it offered an overview for cultural formula consisting of:

- A discussion of the cultural variants in presently identified DSM disorders.
- A reference of culture-bound syndromes.
- An overview of a culturally appropriate instance solution based on 5 significant locations:

Language recognition, cultural health problem

definitions, psychosocial environment social factors and operating rates, cultural elements of communicating the patient with the medical professionals, and cultural evaluation in general for diagnosis and treatment.

Things that need to be included in thinking about a client's social identity are social referral teams, involvement with culture of beginning, language, and social aspects of growth. For Native Americans, it is crucial to note the people the person is component of and what tribe or ethnic team the person determines with. Another factor that ought to be taken into consideration is whether the person talks about his indigenous language and what his first language was. Typically, people can feel pushed away from their host society if they do not talk about their indigenous language with complete confidence or whatsoever. This can be an obstacle to those wishing to look for treatment from standard therapists. It is likewise essential to note what involvement a person has actually had with his host culture and to what degree his household is involved with their society.

An instance of social alienation was seen in previous generations of Native American youngsters who were sent to boarding schools. This, at some point, led to an increased weakening of the society that had actually formerly led Native American communities. Lots of Native Americans believe that this loss of culture is the primary cause of their existing social troubles, which includes those connected with alcohol.

Screening can be done either consistently by asking individuals about their alcohol and drug use in the past year or by utilizing a screening examination, such as the Alcohol and also Use Disorders Identification Test (AUDIT) to identify the amount of alcohol use, or the Michigan Alcohol Screening Test (MAST), which has actually been customized to include the usage of drugs. One research that utilized the brief variation MAST found that it may have produced a high number of incorrect positive when utilizing the cutoff rating of 3 or greater.

Treatment

One study showed that a particular ethnic group was not looking for alcohol and medicine from a neighborhood program because the program had no staff from participants of the same ethnic group20. The composition of the staff is critical to establish therapy and retention programs in particular. Moreover, it's best for the treatment provider to take on an inquisitive role and make no ethnic-centered assumptions based on its own cultural heritage unless the treatment provider has the exact same ethnic history.

Accessibility to treatment is promoted by finding therapy centers in quickly obtainable geographical areas. Patients ought to have access to centers and also therapists in their own area rather than in remote therapy locations.

One facet of healing that is usually neglected is that of cultural healing. Cultural recovery entails restoring a sensible ethnic identity and also obtaining a practical social network dedicated to the individual's recuperation; making a religious, spiritual, or moral recommitment; re-engaging in recreational or trade activities; as well as acquiring a social duty in the recouping community, society at large, or both — those people who fail to make a satisfactory cultural recovery go to danger for re-addiction.

Household participation is an important focus in collaborating with Hispanic as well as Native American communities. Both the patient's immediate family members and expanded family members are substantial and also must be associated with the treatment procedure because alcohol and drug abuse can erode vital household as well as social ties, and also restorative initiatives to repair a person's social and familial network can buffer the impacts of alcohol or drug abuse.

Finally, the neighborhood needs to redevelop a culturally incorporated textile, just component of which might be related to alcohol and drug use. Initiatives to redevelop a culturally integrated community needs to precede, or at least parallel, the growth of a significant intervention; efforts must incorporate fundamental neighborhood social worths with the most current developments in therapeutic intervention. For example, the Alkali Lake area in British Columbia accomplished a reduction of alcoholism from 95% to 5% over 10 years through the revitalization of practice as well as the establishment of a community environment that no longer tolerated alcohol addiction. As Chief Andy Chelsey basically it "The area is the therapy center."

PSYCHOACTIVE SUBSTANCE

There have been countless interpretations of culture. It offers as a guide for activity, a cognitive map, and a grammar for habits."

Substance abuse describes the misuse of alcohol and also other drugs, mainly illicit drugs, but what is thought about "illegal" is often culturally determined and can vary between social teams. The majority of culturally distinctive groups have actually made use of and also overused alcohol and also various other drugs throughout the ages, and they have actually established a code of conduct in their approach to alcohol and drugs.

This short book begins with a brief review of the historical background as well as context for the usage and also misuse of substances. A testimonial of the effect of society on the initiation, usage, and misuse of substances adheres to.

Roots, Fruits, and Plants Historic

Alcohol, as well as lots of various other medications, have been used for thousands of years. Some of the earliest references to the use of alcohol are discovered in old Sumerian clay tablet computers that consist of dishes for the use of red wine as a solvent for medications. There are some isolated records of alcohol usage by the Aztecs in Mexico, by the Pima/Papago in the Southwest United States, as well as by the Aleuts from as far north as Alaska.

At regarding the same time that some teams were fermenting liquors, the Sumerians were growing the opium poppy, which they called "hulgil," the plant of joy. The opium poppy was used to relieve pain and relaxation of the intestines for its residential medical properties and also to provide sedation and euphoria for its mental health properties. The presence of the opium in the United States was believed by early Chinese immigrants, and heroin was later introduced to minority groups in metropolitan areas, such as the Black and also the Hispanics.

It is claimed that marijuana originated in China and later in India more than 4000 years ago. Before using cannabis as a fiber, the examples of its cloth usage were used in China over 10,000 years earlier.5 Hannover was developed at Mount Vernon by George Washington and was the second-largest plant— after cotton— cultivated in the Southern region before the U.S. Civil War. After the First World War, Mexican workers welcomed Americans to their psychedelic buildings to smoking marijuana. Upon arrival in the Caribbean, Christopher Columbus encountered the locals who use tobacco for a variety of illnesses. Peyote was used in northern Mexico's ceremonial rituals and later extended to native American tribes in the Southwest.

THE ROLE OF CULTURE IN SUBSTANCE USE AND ABUSE

One reason is that the ancient Aztecs used alcohol before white settlers called any kind of call. They were very controlled to use alcohol and were for ceremonial purposes only. Peyote was used for chronic alcohol addiction in a ritual environment.

Initiation right into extreme substance usage might occur during durations of rapid social modification, frequently amongst social teams that have had little exposure to medicine as well as have not created protective normative behavior. Anomie, or loss of a healthy ethnic or cultural identity, might happen among native populaces whose cultures have been devastated by the comprehensive and abrupt increase of outside influence.

A couple of North American Indians had any direct exposure to alcohol before the arrival of whites since of its low accessibility. On the western frontier, powerful distilled alcohols came to be commonly offered, as well as the only model Native Americans had was the drunken comportment of the frontiersman.

Acculturation, the level to which a private understands his/her native society, is thought to be related to substance usage and misuse. Native American seniors believe that many drug abuse problems relate to the loss of typical culture. Higher rates of important usage have actually been discovered in persons that closely identify with non-Native American worths, and also, the most affordable prices are found in bicultural people who are comfortable with both sets of cultural worths.

Immigrants leave the protective setting of their household behind and are encountered with a new collection of cultural standards as well as values. These ladies may commonly assume the alcohol consumption habits of the leading culture as well as, as a result, they enhance their use of alcohol.

A current study from Washington State demonstrated the accelerated misuse of alcohol and also used controlled substances in acculturated Hispanics.15 Illegal medicine use in the previous month and boosted alcohol usage was reported by 7.2% compared with less than 1% of non-acculturated Hispanics and also 6.4% of whites. Non-acculturated Hispanics (recent immigrants) were even more, household oriented as well as had reduced prices of alcohol and drug use. Obviously, native social worths have a safety effect.

ASSESSMENT AND SCREENING

With the population of the United States coming to be progressively varied, it is important to consider an individual's social background when analyzing for drug abuse or dependence. The magazine of DSM-IV was an important turning point in the application of social psychiatry concepts since it supplied an overview for social formula including:

- A discussion of the cultural variations in presently recognized DSM disorders.
- A reference of culture-bound syndromes.
- A contour of a culturally relevant example solution based on five main areas:
  social recognition, social interpretation of the disorder, social factors related to the psychosocial setting and organizational rates, social aspects of the patient and clinical relationship, and general social framework for health diagnosis and treatment.

The things that ought to be included in taking into consideration a patient's cultural identity are social referral teams, involvement with culture of beginning, language, and also cultural factors of growth. For Native Americans, it is crucial to keep in mind the tribe the individual is composed of and also what people or ethnic team the person determines with. An additional factor that must be considered is whether the person speaks his indigenous language and also what his mother tongue was. Frequently, individuals can feel pushed away from their host society if they do not speak their native language with complete confidence or in any way. This can be an obstacle to those wishing to look for treatment from traditional therapists. It is also essential to note what involvement an individual has actually had with his host society and to what degree his family is included with their society.

An example of cultural alienation was seen in previous generations of Native American kids who were sent out to boarding institutions. This, at some point, led to an increased weakening of the culture that had actually formerly assisted Native American areas. Many Native Americans think that this loss of culture is the primary reason for their existing social issues, which consists of those associated with alcohol.

Testing can be performed on a regular basis, either through the previous year's request of patients with regard to their alcohol use and also the use of drugs, or by the use of a screening test, such as alcohol or AUDIT (AUDIT) to measure the amount of alcohol use, or the customized Michigan Alcohol Screening Test (MAST) for drugs. Research using the short variance MAST found that the use of the cutoff rate of 3 or more could have created a significant number of false positives.

Therapy

One research found that a certain ethnic team did not look for alcohol or drug treatment from a neighborhood program since the program did not have a team that consisted of participants of the same ethnic team.[20] Staff composition is essential in developing therapy programs, especially with treatment initiation, as well as retention. In enhancement, if the treatment carrier is not of the very same ethnic background, it is finest that he or she takes on a curious duty and also not make any kind of ethnocentric assumptions based on his very own cultural heritage.

Accessibility to therapy is facilitated by finding treatment facilities in conveniently accessible geographic locations. Patients should have access to facilities as well as therapists in their own neighborhood as opposed to in remote treatment places. One caveat is that in tiny country neighborhoods, simplicity of gain access might decreases the capability to maintain treatment personally. This is largely based on whether the person who supplies therapy lives within the local community or outside of it.

One element of recuperation that is commonly neglected is that of social recovery. Cultural recovery includes regaining a viable ethnic identity as well as obtaining a functional social media network dedicated to the individual's recuperation; making a spiritual, spiritual, or ethical recommitment; re-engaging in a trade or leisure activities; and also acquiring a social role in the recuperating community, culture at huge, or both. Those individuals that stop working to make sufficient cultural recuperation are at danger for re-addiction.

Family member's participation is an important focus

in working with Native and Hispanic American communities. Both the client's prompt household and prolonged household are significant as well as need to be involved in the intervention process because alcohol and substance abuse can erode essential family members as well as social ties, and restorative efforts to repair a person's familial and social network can buffer the results of alcohol or substance abuse.

The area should redevelop a culturally incorporated fabric, only part of which may be connected to medicine as well as alcohol use. Efforts to re-establish a culturally integrated area needs to precede, or at least parallel, the growth of a significant treatment; initiatives should incorporate fundamental community cultural worths with the most current developments in treatment intervention.

HALLUCINOGENS

Hallucinogens are drugs that change the user's thinking processes as well as the assumption in a way that results in considerable distortions of fact. These medicines impact one's understanding much in a different way than several other kinds of drugs do. To many, the impact of these medicines stands for experiences of brand-new as well as also expanded consciousness, and, certainly, some individuals experience synesthesia (combined sensory experiences, such as seeing audios or hearing shades). Other usual impacts produced by these drugs include hallucinations, a transformed sense of time, as well as dissociative experiences (e.g., not really feeling linked to one's body or fact).

A lot of hallucinogens are identified by the United States Drug Enforcement Administration (DEA) as Schedule I managed substances, meaning they have no recognized medical uses as well as have a high capacity for abuse and also psychological or physical dependency. Thousands of compounds are classified as hallucinogens.

Some of the usual extra hallucinogens consist of:
- LSD
- Psilocybin (magic mushrooms)
- peyote (mescaline)
- DMT.
- Ketamine (Special K)
- PCP (phencyclidine)

Ketamine is a Schedule III medication, and also PCP is a Schedule II, due to their previous medical usages. However, they are severe medicines of issue.

While these are not usually essential abuse drugs, approximately 1,2 million people aged twelve years and over reported hallucinogens in 2014. We are used for a variety of reasons consisting of religion, stress relief, entertainment, or to achieve understanding or enlightenment. Such medications may have serious effects and impairments, although not classically addictive.

Kinds of Hallucinogens

Hallucinogens can be classified into two subcategories: the classic hallucinogens and the dissociative medicines. Timeless hallucinogens normally generate aesthetic as well as auditory hallucinations as well as may lead to a transformed sense of time as well as enhanced sensory experiences. Dissociative medicines produce sensations of detachment, such as derealization (the feeling that is removed from fact or that things are unreal) as well as depersonalization (the feeling that one is separated from one's own physique).

The classic hallucinogens talked about in this book consist of:

- LSD
- Psilocybin
- Peyote
- DMT
- LSD

The fungi had no sensible usages for this objective, and also it was shelved. Five years later on, Hoffman started working with it again, and also, after mistakenly absorbing it via his fingertips, he experienced the medication's hallucinogenic effects. The majority of individuals using LSD usually feel euphoric, experience aesthetic hallucinations, and often have a really extreme state of mind;

nonetheless, so-called "bad trips" can take place in people, resulting in extreme stress and anxiety (consisting of panic attacks) and also considerable anxiety.

LSD is typically taken as a capsule liquid, or "blotter paper" that has been dosed with LSD liquid. A typical dose average in the plain micro-milligram range, the results can last up to 12 hours. LSD was an extremely prominent medication in the 1960s as well as the very early 1970s, as well as its use, was partly accountable for the medication culture of that time.

LSD usage does not appear to lead to the physical reliance, although tolerance can create. Other possible impacts of LSD consist of:

Increased body temperature, heart rate, as well as blood pressure

- Profound sweating
- Dizziness
- Loss of cravings
- Dry mouth
- Tremors
- Numbness
- Impulsiveness
- Mood swings
- Hallucinations
- Distorted reasoning

Long-lasting LSD usage, in uncommon cases, can

result in Hallucinogen Persisting Perception Disorder, or persistent flashbacks of experiences while on LSD 5. These flashbacks can create substantial impairment or distress in the customer's life and also can last for several years 5.

Psilocybin

Psilocybin (4- phosphoryloxy -N, N- dimethyltryptamine) is a hallucinogenic substance that is located in greater than 200 sorts of mushrooms.

In particular areas in South America, Mexico, and the United States, these champignons are generally found. Typical mushroom street names with psilocybin include enchanted mushrooms, champagne, and shrooms. The champignons are usually eaten and are also sometimes polished as a drink.

A couple of results of eating mushrooms consist of:

- Relaxation
- Spiritual experiences
- Hallucinations
- Panic
- Paranoia
- Psychosis
- Nausea
- Vomiting

One threat related to psilocybin use is that of

poisoning. Individuals might misidentify the mushrooms and accidentally ingest toxic mushrooms, which can cause fatality.

Peyote

It was used in Mexico by the Aztecs as well as by certain teams of Native Americans. These teams used it for medicinal and also hallucinogenic purposes. Some Native American churches still have the legal right to make use of peyote in religious solutions in spite of its category by the DEA.

Generally, the peyote cactus buttons are eaten or taken in water. They can likewise be ground and placed in a capsule or smoked with tobacco or marijuana.

Customers tend to experience the psychoactive impacts of mescaline within one to two hours after ingestion, as well as its impacts can last approximately 12 hours. These impacts consist of:

- Increased heart rate and body temperature level
- Vomiting
- Flushed skin
- Extreme sweating
- Coordination problems
- Hallucinations
- Altered understanding as well as body image
- Anxiety

It's not most likely that individuals utilizing peyote or mescaline will certainly end up being addicted, however resistance, in addition to cross-tolerance to other hallucinogens, can create. Routine usage does not show up to result in the development of physical reliance, and also withdrawal signs and symptoms are unusual.

DMT

DMT (N,N-Dimethyltryptamine), or "Dimitri," is a hallucinogenic chemical that occurs naturally in some Amazonian plants, but can additionally be artificially manufactured. When made in a research laboratory, DMT resembles a white, crystalline powder and is most often smoked. The around the world use of DMT is rising, as it has a multitude of new individuals compared to other drugs. Small amounts of DMT may take place normally in the human brain. These trace quantities of DMT are assumed to be associated with people's records of specific unusual events, such as near-death experiences, mystical experiences, or unusual abductions.

Unlike many various other hallucinogens with a relatively long period of time of results, DMT creates an extreme yet temporary intoxication. Overall, customers have not reported many unfavorable adverse or "comedown" impacts.

The impacts of DMT may consist of:
- Hallucinations
- Body and spatial distortions
- Changes in awareness and understanding
- Increased heart price and also blood pressure
- Agitation
- Severe throwing up (as a result of ayahuasca tea)

Long-lasting DMT usage does not appear to create tolerance, and there is little proof bordering the long-lasting results of ayahuasca usage. The worldwide use of DMT is boosting, as it has a big number of brand-new customers compared to other drugs.

Unlike several various other hallucinogens with a fairly long duration of effects, DMT produces a temporary but extreme intoxication. Overall, individuals have not reported many negative, damaging, or "comedown" effects.

PCP
Phencyclidine (PCP) was originally established as an anesthetic; however, because its use is connected with major side effects, the dissociative medication is no longer made use of medicinally. It's still lawful for usage in animals yet is rarely made use of in vet settings. Pure PCP is white and also crystalline in look; however, additives might offer it a tan or brown color. PCP is typically taken orally in a tablet or pill type, smoked, grunted as a powder, or injected. Street

names for PCP consist of angel dust, animal depressant, and also rocket fuel. A minimum of 14 types of PCP was offered on the street between the late 1960s and 1990s and also many illegal examples have PCC, a poisonous chemical that launches cyanide as well as can create poisoning. The occurrence of PCP addiction or PCP make use of condition is unknown, concerning 2.5% of the populace has reported utilizing PCP at least once in their lives.

The results of PCP differ depending upon the dosage, yet in general, the customer will certainly really feel impacts within 1-5 minutes if the hallucinogen is injected or smoked and also within about 30 mins if taken orally or snorted.

Intoxication normally lasts concerning 4-6 hrs, as well as impacts might include:

- Euphoria
- Feelings of invulnerability and stamina
- Disorientation
- Distorted sensory assumption
- Disordered thoughts
- Hallucinations and illusions
- Bizarre or fierce behaviors
- Severe anxiousness
- Amnesia
- Paranoia

- Numbness or reduced feedback to pain
- Seizures

The effects of the medication are often boosted when PCP is mixed with other substances, such as alcohol, stimulants such as drugs, or downers, including numbing medications. Mixing PCP with alcohol or other medications can increase the risk of unfavorable effects and also overdose.

PCP drunkenness raises the danger of injuries from attacks, accidents, or drops 5. Chronic PCP use can result in disabilities in memory, cognition, as well as speech, as well as these shortages, which may last for a month.

It's not uncommon for long-lasting PCP customers to also experience:

- Heart strikes
- Respiratory issues
- Intracranial hemorrhage (blood loss inside the head)
- Rhabdomyolysis (the breakdown of muscle mass cells, which can bring about kidney failure)
- Depression

Chronic PCP customers may create resistance and need greater doses of the medication in order to experience desired effects 5. This can be unsafe since greater dosages can cause seizures and coma.

Ketamine

Ketamine was created as an anesthetic for both

animal and human usage, specifically in trauma or emergency situation situations. Nowadays, ketamine is abused for its dissociative impacts as well as its popularity as a "club medication" is boosting, particularly among young people and teenagers.

The drug can be grunted, smoked, injected, or mixed right into drinks. It is often utilized in combination with cocaine, amphetamine, MDMA, or methamphetamine (Ecstasy).

Road names for ketamine consist of:
- Special K
- K
- Cat depressant
- Kit kat

The effects of ketamine take place quickly and may include:
- Sedation
- Numbness
- Hallucinations
- Delirium
- Psychosis
- Paranoia
- Disorientation
- Feelings of detachment
- Depression
- Agitation
- Amnesia

- Cognitive impairments
- Nausea
- Muscle tightness
- Heart palpitations
- Dizziness
- Seizures

Tolerance to ketamine use establishes swiftly as well as there is proof of physical dependence in persistent individuals. There are documented situations of withdrawal symptoms in some people, but an inadequate research study exists to sustain a ketamine withdrawal disorder.

Are Hallucinogens Addictive?
Lots of people correspond to the term dependency with the experience of withdrawal symptoms, despite the fact that both are different issues. Hallucinogen customers do not have a tendency to experience withdrawal symptoms with the cessation of usage due to the reality that these drugs do not have a high capacity for physical dependence. As well as although hallucinogens aren't typically addicting, individuals can still deal with troublesome use that hinders their everyday lives. As a result of the uncertainty bordering the term "addiction," it is no longer used scientifically in the diagnostic procedure.

Instead, the term compound use condition is utilized

to indicate a psychiatric/psychological problem that happens in people that experience unfavorable implications and also problems regulating making use of medicines. The American Psychiatric Association lists certain analysis standards for a hallucinogen usage disorder, which includes both hallucinogen abuse and unfavorable repercussions of usage.

People who make use of these substances for non-medicinal functions, have problems regulating their usage, and experience adverse effects as an outcome of their use might be diagnosed with a hallucinogen usage condition or phencyclidine make use of disorder in the case of PCP usage.

MARIJUANA

Cannabis describes the dried leaves, stems, flowers, as well as seeds from the hemp plant Cannabis. Chemical delta-9-tetrahydrocannabinol (THC) is the main active ingredient in marijuana.

The most widely used controlled substance in the United States is cannabis. In 2017, 45% of Americans over the age of 12 used marijuana, according to a nationwide study on drug abuse or well-being.

In addition, DC had federal law on cannabis for legal use by adults aged 21, since the 2018 medium-term elections of 10 states and Washington. More than 30 states have only books that legalize marijuana for medicinal use, while many others only have legalized oils with low THC content. The following is also known: More than 200 terms of jargon are available for marijuana: pot, natural herb, weed, grass, widow, boom, hashish, hash, Mary Jane, Cannabis, periodontal bubble, northern lights, fruity juice, sweet, skunk, and chronic. Cannabis is still illegal under federal laws.

Medication Class: Marijuana is frequently identified as a depressant, although it also has energizer and hallucinogenic residential or commercial properties.

Usual Side Effects: Side effects of cannabis use consist of transformed detect, state of mind changes, problem assuming, and also damaged memory. In high doses, it can lead to hallucinations, psychosis, and also misconceptions.

Exactly How to Recognize Marijuana

Cannabis looks like a shredded, green-brown mix of plant material. However, it can look various, relying on how it is prepared or packaged.

What Does Marijuana Do?

The membranes of different nerve cells in the brain are composed of receptors that bind to THC, causing a series of cell reactions, which ultimately lead to a high level of cannabis use. Smoking marijuana is the most typical way of using it.

A recently preferred technique of use is smoking or

eating various kinds of THC-rich materials removed from the marijuana plant. It can additionally be baked into food (called edibles) such as brownies, cookies, or sweet, or brewed as a tea.

What the Experts Say
Marijuana use can be particularly troublesome amongst teens since it might have a long-term influence on mental abilities consisting of memory, learning, and also thinking. One 2012 research discovered that individuals that had actually started smoking cigarettes cannabis in their teens lost approximately 8 INTELLIGENCE points.

Due to the fact that the most common technique of use is smoking cigarettes, cannabis usage additionally poses breathing threats and various other smoking-related dangers. Smoking marijuana might increase the threat of hissing, lack of breath, as well as persistent coughing. According to an evaluation released in 2015, the study is mixed on whether smoking cigarettes marijuana raises the danger of cancer cells. Some researches have actually suggested that there might be a raised risk, while others have found that cannabis use might really have a safety result.

Despite these threats, there are reasons why individuals pick to remain to utilize cannabis.
One research study published in 2016 discovered people reports using marijuana to:

- Relieve stress and anxiety or tension
- Escape life's troubles
- Ease dullness
- Feel excellent or euphoric
- Fit in socially
- Off-Label or Recently Approved Uses

Along with its use as an entertainment medication, marijuana has a lengthy history of usage for medicinal objectives. While it has not been approved by the FDA, several states in the U.S. have legalized marijuana for at least some clinical purposes.

Clinical marijuana is made use of to deal with the signs and symptoms of conditions rather than as a treatment for the condition itself. Research through 2017 suggests that cannabis is most effective in the treatment of muscle mass spasms, persistent discomfort, and nausea, making it practical in alleviating the symptoms of conditions such as numerous sclerosis (MS) and also epilepsy.

Some of the problems that clinical marijuana has been authorized to treat in numerous states include:

- AIDS
- Alzheimer's condition
- Cancer
- Crohn's condition
- Eating problems

- Glaucoma
- Cachexia
- Migraines
- Seizures
- Severe pain
- Severe queasiness
- Persistent muscle mass spasms
- Wasting disorder

Additional study on the possible advantages of clinical cannabis is ongoing. Acknowledged and also legally sanctioned use cannabis for the treatment or alleviation of signs and symptoms will continue to progress as scientists check out these usages.

Since 2019, medical marijuana is lawful in 33 states as well as Washington, D.C.

Common Side Effects

Some of the typical adverse effects of using marijuana are a totally dry body, puffy eyelids, bladder eyes, sync loss, and accelerated cardio activity.

Short-term dangers include:
- Anxiety and also paranoia
- Impaired memory
- Difficulty reasoning
- Learning problems
- Lack of attention and also focus

- Poor driving skills

Lasting threats possibly include:
- Respiratory issues
- Heightened risk of infections, specifically the lungs
- Poor short-term recall
- Cognitive impairment
- Lack of inspiration

Normal cannabis cigarette smokers might likewise have much of the exact same respiratory system troubles that cigarette smokers have, including everyday coughing and phlegm, signs and symptoms of persistent respiratory disease, and a lot more regular upper body colds. Continuing to smoke marijuana can cause the abnormal performance of lung cells injured or ruined by cannabis smoke.

While a few of these threats can't be alleviated, there are things you can do to deal with-- at least partially, a few of the above, if you pick to smoke.

Indicators of Use

Marijuana can be consumed in a variety of means, although smoking cigarettes is one of the most common methods.

If you think that a person you recognize is misusing cannabis recreationally, you may see a few of the following indications:
- Lack of empathy

- Talkativeness
- Secrecy
- Sleepiness
- Increased food cravings
- Bloodshot eyes
- Poor time monitoring
- Drug materiel (e.g., pipes, baggies, rolling documents)

It is necessary to bear in mind that much of these indicators may be triggered by other points or might just be variants in regular actions. Look for teams of habits as opposed to taking single actions as proof of drug use.

Myths and Common Questions

One typical misconception is that marijuana is a portal medication that often brings about using "more challenging" drugs. While there is some evidence that exposure to cannabis might make it simpler to make use of various other substances, the National Institute on Drug Abuse (NIDA) recommends that many people who utilize cannabis do not take place to try to end up being addicted to other medicines.

One more typical misconception is that marijuana itself is not habit-forming. While it is not typical, repeated usage can lead to both mental and physical reliance. The Centers for Disease Control reports that as many as 1 in 10 individuals who make use of marijuana will certainly establish an addiction. Marijuana today commonly contains a lot greater THC levels than in the past, which raises its habit of forming buildings.

Tolerance, withdrawal, and also dependancy
Research suggests that regular use of cannabis may lead to resistance. In a 2018 study, scientists found that routine usage of marijuana led to much less popular effects when contrasted to non-regular use.

How Much Time Does Marijuana Stay In Your System?
The quantity of time cannabis stays in your system may depend on the dosage as well as the frequency of use. Typically, marijuana may be discovered in urine examinations for up to 13 days after usage. However, regular usage might lead to longer detection of home windows.

Dependency
Research published in 2015 showed that over 30% of cannabis users in the United States had use problem in 2012 as well as 2013. Long-lasting marijuana individuals are more at risk of addiction. Individuals that start utilizing marijuana prior to age 18 are four to 7 times more likely than grownups ages 22-- 26 to create a dependency.

Drug food craving and withdrawal signs and

symptoms can make it hard for long-lasting cannabis smokers to quit utilizing the medication. Individuals are attempting to give up record stress and anxiety, sleep loss, and irritation. Medicine is taken into consideration addicting if it causes a person to compulsively, and typically frantically, long for, seek, as well as use it, even when faced with unfavorable health as well as social effects. Cannabis meets this standard.

Withdrawal
A few of the common signs and symptoms of cannabis withdrawal that individuals report experiencing consist of:
- Difficulty sleeping
- Drug desires
- Decreased hunger
- Anxiety
- Irritability
- Mood modifications
- Headaches
- Chills and also sweats
- Depression

These symptoms may vary between mild and extra severe. Usually, these symptoms of withdrawal can be self-managed, but if they are serious, longer, or psychiatric depression, you ought to speak with your medical professional.

How can I get assistance?

Marijuana therapy often uses therapy as well as psychiatric treatment. The goal is to help people learn new behaviors and overcome existing addictions and psychiatric problems co-occurring.

Types of counseling or treatment that might work include:

- Cognitive-behavioral treatment
- Motivational rewards
- Individual or team counseling
- Family therapy
- Support teams.

Although the therapy of the marijuana problem has not been approved, antidepressants and other medicines may be used to cope with signs and symptoms of problems such as depression or anxiety.

PSYCHOACTIVE SUBSTANCES: HARM REDUCTION POSITION STATEMENT

Many factors were dealt with promptly at the recent special meeting of the UN General Assembly on drugs. Obviously, the' drug war' has simply been a short-a single focus on reducing supply does not succeed.

The legal status of alcohol, cigarettes, marijuana, and other medicinal drugs or pharmaceutical products has changed globally with countries trying to develop effective supply-reducing policies, reducing the demands and decreasing accidents.

There is raising emphasis on the relevance of carrying out evidence-based policies to address the considerable morbidity and also death related to making use of alcohol, cigarette, marijuana and various other psychedelic substances.

Psychedelic substances are subject to a number of worldwide and also national conventions and legislations that have stressed the relevance of the 'war on medications' as well as supply reduction. The Prevention of and Treatment for Substance Abuse Act 70 of 2008 talks to the National Drug Master

Plan, which stresses a range of methods for dealing with the too much use of alcohol, cigarette, marijuana as well as other psychedelic substances.

Supply reduction describes policing efforts to curb the manufacture as well as the distribution of alcohol, cigarette, cannabis, and other psychedelic substances or medications.

Need decrease refers to precautionary efforts to lower their demand. Damage decrease describes treatments and policies to reduce the hazardous effects of alcohol, tobacco, marijuana as well as other psychoactive substance use. Concentrating on injury decrease does not suggest that dangerous behaviors as a whole, and using psychoactive substances or medications specifically, rate.

This method is based rather on the clinical evidence on what works to boost public health and wellness and also decrease social injuries when tobacco, alcohol, cannabis as well as other psychedelic substances are already being used.

While alcoholics, cigarettes, marijuana, and other psychedelic drugs are associated with a wide range of possible damage to people and society, not all pseudoactive materials are harmful.

In terms of its relationship to physical disease, psychological disorder, social violence and crime, alcohol is obviously the most dangerous compound used for South Africa, based on global experience.

Growing proof shows that details plans and

interventions can reduce the potential injuries associated with the continued usage of psychoactive substances.

Other efficacious interventions for lowering harm in people with alcohol, cigarette, cannabis, opioid, and various other medicine usage include medication-assisted therapy as well as needle and also syringe programs.

9 THINGS THAT MATTER ABOUT PSYCHOACTIVE DRUGS

Psychedelic medications chemically alter the brain as well as change the means we really feel, think, regard, and comprehend our globe. They are common: alcohol, cannabis, opioids, cigarette, sedatives, hallucinogens, and also energizers, among others. Some happen naturally-- nature's payment to our psyches and bodies-- and some are manufactured in laboratories to influence the very same mind receptors as do those discovered in woodlands, deserts as well as open fields. We are in a psychedelic medicine epidemic in our nation, most notably the opioids, due to their awful death toll.

We need solutions to the epidemic to conserve communities, lives as well as families-- and federal government treasuries. However, if we focus just on the medicine itself, whatever it might be, we will miss what truly matters when it comes to just how human beings reply to psychedelic representatives.

Right here are nine things that matter when it concerns drugs:

1. Age

That's since the human mind is still under building and construction till well into the 20s, later on for males than females. It takes virtually three years for the brain to lay down the fatty compound totally, myelin that surrounds the nerve connections and permits reflection and controls impulsive activity, for the cortex to stand a chance versus the drive focuses deeper in the brain. Repetitive or high dosages of psychoactive drugs like cannabis, alcohol, and also hallucinogens conflict with the typical development of the brain.

Cigarette smokers are much more common among teenagers who smoke while younger than 18. Early drinking is a sign of biological vulnerability to alcohol at age 12, 13, or older. Therefore, when a person is living in the seventh and eighth decades, the aging brain is very susceptible to psychedelic drugs, and also small amounts look similar in large numbers.

2. Set and Setting

Establishing implies that the individual is especially human, physiological, experiential, and emotional. The set creates an individual vulnerability and also selective responsivity to substances. Of example, the influence of psychedelic drugs can be influenced by a person's genes (inherited DNA) as well as by current mental neurochemistry or whole-body physiology.

The exact same amount of medication might have orders of size, basically, in its influence. Additionally, repeated use of material can produce the main nerve system hyper- or hypo-reactivity to that representative.

Psychologically and also experientially, a background of injury (from abuse, neglect, physical violence, and abuse, compelled migration, and also an all-natural disaster) generates fantastic mind (and also emotional) sensitivity to several things, consisting of medicines.

Unstable elements of an individuality

specifically a tendency to externalize, to hold others in charge of whatever, along with measurements of individuality, such as passive or active, rebellious or adapting, the capability to experience sensations or not, and also denying or approving fact-- all influence the action of a material.

A historic tale reveals what setting ways. At the height of the Vietnam War, the Department of Defense pertained to understand that 20 percent of the soldiers were regular customers of the powerful heroin they had easy access to. The DoD feared that when they returned, the people in the U.S. who are heroin-addicted were to join them.

Norman Zinberg (a former coworker, now deceased), as well as his associate Lee Robins, were sent off to Vietnam to attempt as well as examine the problem to forecast the future for these soldiers. They were proven to be appropriate when they forecast a no better price of heroin use or reliance than existed in those that did not go to war.It was the soldiers' setting-- fighting in a fierce, harmful, unpredictable guerrilla battle in a country that did not desire them and with little assistance from Americans back home, with all set accessibility to economic, powerful heroin to make the intolerable acceptable-- that brought about their high prices of use.

Today's analog, so partial, is Iraq and also Afghanistan, where research studies reveal that as a number of 30 percent of fight experts return with PTSD, anxiety, or stressful brain injury (TBI). Their setting is generating high rates of alcohol addiction and substance abuse.

Establishing an individual is in the issue.

3. Path of Administration

Intravenous (or intraarterial) pushing on a syringe as well as a bam in the brain is believed by the majority of individuals to be the fastest route to the brain. A venous charge of a drug such as heroin first has to work its way back to the heart, then go to the lungs via the lung artery and to the mind through the carotid artery. It's easy for sure, but when a drug is inhaled, the first vein moves from the body to the heart.

Our minds offer leading concern to getting oxygen. A drug that can accompany oxygen, right from the lungs, will certainly be the first to show up and also do its job.Some believe this is the factor that cigarettes (or vapes) are the most habit-forming of substances, more challenging to quit than heroin.

If you desire quickly and also angry, and you're not speaking about an action flick, breathe in deeply, and also you are there.

4. Pureness

It makes both common as well as a medicinal sense of how pure a compound makes a huge difference. Heroin is typically combined with baby powder, strychnine, or other chemicals that lower its pureness, which takes place at about every step of the supply chain to boost revenue-- from the watercraft that smuggles it into the gangs that distribute it to the regional, regional and also street dealers. It is why people who are dependent on high dosages of heroin typically discover a solution inadequate also to abate their withdrawal, a lot less obtain them high.

The exact same, obviously, relates to cocaine as well as methedrine.

This is in part why fatalities from opioid overdose keep expanding: suppliers are blending it with fentanyl (and its relatives) to offer the customer a much more powerful hit. Pure heroin, or morphine, is much more secure, as well as why some nations have adopted the injury reduction strategy of making these opioids lawful.

The purer the drug, the less its contaminants, the higher its influence on our main nerve system.

5. Effectiveness

The more powerful the medication, the much more it takes us down its neurochemical journey. A host of different psychoactive substances (consisting of heroin, cocaine, meth, even alcohol-- the least understood and a lot of made use of material around) create the release of dopamine, especially in a section of the mind called the center accumbens, which is a important however little nerve complicated deep in the mind, involved in enjoyment, incentive, and aversion. The more potent the medicine, the higher its kick.

Strength counts. We rarely ever know the potency of medicine acquired on the street or the dark web.

6. Half-Life

Take into consideration Xanax (alprazolam), in some cases called Vitamin X, which was introduced for public usage in the U.S. in 1969. I made use of to suggest it, specifically for individuals with disabling anxiety or sleeping disorders pertaining to clinical depression, for the numerous weeks it considered the antidepressant I had actually also recommended began to function. However, I soon found out that its rapid and also efficient action began to fade in an issue of hours, leaving people feeling dreadful or awakening during the night, yearning even more of the medication. That's why it has actually largely been changed by longer-acting representatives.

The half-life of medicine is the moment it takes for the blood degree of the compound to reduce by 50 percent. Xanax's half-life was formally 11 hrs, usually, but clients informed me they might feel it subsiding a lot more quickly than that. They might feel it diminishing, reaching its half-life, in a few hours. A solitary drink of an alcoholic beverage taken by an adult has a half-life of concerning half an hr. Methedrine has a half-life ranging from 6 to 12 or more hrs, depending upon the individual.

A drug's half-life, as a result, influences the moment to a craving for even more of its wanted activity, as well as how soon a person really feels the pangs of withdrawal.

7. The Original Source

When a drug is derived from a plant instead of manufactured in a laboratory, the make-up of the plant affects its psychoactive properties, use, and also the capacity for dependence. Cannabis is a fine example. It contains well over 60 cannabinoids, the energetic components. Both major ones are tetrahydrocannabinol (THC) as well as cannabidiol (CBD). THC obtains us high. However, CBD has no psychedelic homes.

Today, many thanks to sophisticated hereditary

adjustment of the plants, THC's potency is typically 60 times greater. That can be mitigated if CBD focus is also high: CBD serves to secure users-- especially young people, or individuals with susceptibility to psychosis, from experiencing psychotic symptoms (which can linger in some individuals also after the medicine is gone from the blood).

The complexity of plant organisms affects their actions, both desired as well as negative.

8. Refinement as well as Extraction

The coca leaf just gathered and also eaten, is much milder in its effects than a drug, which is fine-tuned to have a higher portion of the plant's active psychedelic active ingredient. Crack is much more refined, creating a lot more powerful drugs.

Similarly, the psychedelic drug mescaline is extracted as well as fine-tuned from a cactus native to Mexico as well as components of Texas. And also, any number of plants can be improved, aged and fermented to generate distilled spirits such as vodka, scotch, scotch, tequila, and more.

What matters is not simply the plant, but how people define it.

9. The Drug/Social Forces Ratio

This speaks to the interplay of the medicine, the individual, and also the social setting in which it is used.

The prescription of opioid agonists-- drugs like buprenorphine and also methadone-- has been shown to minimize regression in people depending on opioids. If a person obtaining medication-assisted therapy (MAT) invests time with people that are still making use of opioids or is damaged with signs regarding opioids on TV, social media as well as in music, these social pressures can lead to relapse.

When LSD initially ended up being prominent in the 1960s, metropolitan, healthcare facility emergency departments on a regular basis saw clients in the middle of a poor trip-- in distress or perhaps panic from frightening hallucinations. Poor trips ultimately ended up being much less regular, because users started to understand that taking medicine in a tranquil setup with the assistance of a skilled overview would avoid them.

With all drugs, we need to value that the setup in which a drug is taken in, and also the customer's assumptions of what will certainly come, can affect the action of a material, for better or for worse.

To put it simply, drugs and their uses are complex. They all include psychoactive ingredients. It's the various other "components"-- that we are; how old we are; where, when and also with whom we take the medicines; exactly how pure or unclean they are; how fast they get to the brain and also long they last-- that determine what the experience will be like.

RECREATIONAL USAGE OF PSYCHOACTIVE SUBSTANCE

Some possibly addictive medications have actually been used by a considerable proportion of the populace on a regular basis, to the factor that they have been taken into consideration staple assets. Caffeine, alcohol, and also pure nicotine, being palatable for their mild psychotropic homes, are examples of extensively eaten medicines. As licit psychedelic drugs, they are made use of mainly by "regular" people, in comparison to illegal "hard drugs," which are typically deemed the district of the deviant.7 Nicotine, alcohol, and also caffeine has penetrated our culture, serving as lorries for social interaction, forming our urban landscape, from the Japanese teahouse to the British club, promoting the opening of international trade paths. In a similar way, hashish (marijuana) has been mostly taken in - eaten and later smoked - in Islamic cultures. All these substances have a lengthy history, elaborately intertwined with myth, attesting to guy's partiality for psychoactive substances. The oldest seeds of cultivated creeping plants thus far found and also carbon dated were discovered in Georgia and also belong to the duration from 7000 to 5000 BC.8

According to Jewish as well as Christian custom, among Noah's first activities after coming out of the Ark was to plant a vineyard; he drank a few of its wine as well as became intoxicated (Genesis 9, 20-21).

At the end of the 15th-century, coffee was used mainly across the Islamic world. The use of coffee plants spread quickly to Asia, as did the people of Europe. The history of tea is much older because in China, in the third century BC, the plant was already cultivated.

These essential products have long been the object of main attention, for the objective of collecting excise tax obligation as opposed to managing abuse. In order to remove earnings, leaders in Ancient Egypt, as well as Babylon, established manufacturing or sales syndicates.9 Ordinances limiting consumption have existed together and also alternated with complimentary supply, in close temporal as well as geographic distance. Temperance activities caused a clear decline in liquor use in Western Europe in the early 20th century, finishing with prohibition in the United States (from 1920 to 1933) and in a couple of Nordic nations. In coming before centuries, cigarettes and also marijuana had actually likewise understood prohibition. Cigarette smokers ran the risk of having their lips cut under the initial Romanov tsar, Michael Fyodorovich, or of being beheaded under the Ottoman sultan Murad IV. In 1378, the

Ottoman emir in Egypt, Loudoun Sheikhouni, was established to stamp out hashish usage: farmers expanding hashish were put behind bars or executed, and those found guilty of consuming were said to have their teeth took out.10.

Entertainment medicines are chemical substances taken for satisfaction, or leisure objectives, rather than for medical reasons. Tobacco, high levels of caffeine, and alcohol can be classified as entertainment medicines but are not covered in this brochure. Entertainment medications are normally started to offer pleasure or improve life in some means.

What are recreational medications, and why are they made use of?

Leisure drugs are chemical substances that are used for pleasure. There are lots of reasons individuals attempt recreational drugs.

These consist of:

- Their close friends are doing it, as well as they do not wish to really feel overlooked, or otherwise cool.
- They get pressurized into trying it.
- They are interested in try out the effects and see what happens when they take medications.

- They may feel medicines give them brand-new experiences or perspectives.
- They make them feel much more relaxed or more confident when relating to others.
- They may really feel medications assist them in forgetting their worries or problems.
- They might feel medications make them really feel better.
- They want to be rebellious.
- They delight in the results.

What are the problems with the use of leisure medicines?

All drugs can have harmful impacts. There are numerous troubles which the usage of medications can create.

- Infections. Human immunodeficiency infection (HIV) and certain types of hepatitis can be passed to others by blood on needles. This can happen via sharing needles, or from needlestick injuries.

- Addiction. The majority of recreational drugs can come to be habit-forming, as well as the individual can be depending on having them routinely.

- Cost. A normal drug routine is pricey; acquiring the cash for it can bring about criminal offense, particularly when the person is addicted as well as can not stop.

- Social problems. People regularly using medicines may behave in different ways. This may create issues with their partnerships, or they may lose their task. Children of people that utilize medicines can be influenced.

Drugs can create individuals to have peculiar behavior. Medications can cause individuals to create depression or anxiousness.

- Overdose. Individuals can come to be very unwell or die from medicine overdose.
- Illegal drugs are commonly not pure, as well as people don't always understand what they are taking.
- Accidents, as well as fights, are most likely after taking medicines.
- Unwanted sexual relations are more likely under the influence of drugs.

The number of people makes use of recreational medications?

The government acquires reports on drug use in the UK each year. These are published routinely. The most recent statistics at the time of creating this leaflet are from 2014.

Surveys revealed that in 2014 around 15 in 100 high school students had attempted medicines at some point. Currently, these numbers are dropping, i.e., fewer schoolchildren are attempting medications than in previous surveys.

There were around 371,279 people who were trouble drug customers in the UK, which is around 9 out of every 1,000 individuals. Around 133,112 people consistently injected medications in the UK.

Which are one of the most common entertainment drugs used?

In the UK, cannabis is the most commonly made use of a recreational drug. Some of the much more common entertainment medications are provided below, along with their nicknames, and approximate prices in 2014-2015.

Amfetamines
Various other names: speed, whizz, sulph, dexies.
Exactly how are they taken? Amfetamines are usually offered in powder kind. This can be snorted up the nose, wrapped in cigarette paper, as well as ingested (' speed bomb'), rubbed right into the gum tissues, blended with drinks, or injected.

What do they do? An amfetamine is an energizer, so it gives you more energy. You can keep partying, dance, helping longer without getting tired. It makes you really feel upbeat as well as excited.

What are the hazardous effects? They can make you overactive, nervous, or anxious. Occasionally they create a severe psychological problem where individuals lose contact with fact and see or hear things that are not really there (psychosis).

How much do they set you back? The road cost of amfetamines is around £ 13 per gram.

Amyl nitrates
Other names: poppers, TNT, amyls, kix, liquid gold.

How are they taken? They generally come in a bottle of liquid that is sniffed. They can additionally be breathed in through a cigarette dipped in the bottle.

What do they do? They give you a 'high.' The effects come and go very swiftly. Some people assume they make sex much better.
The liquid is extremely flammable, so it can cause fires if utilized thoughtlessly. They can make you really feel weak or sick.

How much do they cost? The road price is around £ 5 per bottle.

Marijuana

Various other names: hash, hashish, weed, pot, marijuana, marijuana, dope, skunk, yard, puff.

Exactly how is it taken? It can additionally be blended in with food or beverages.

What does it do? It can make you feel happy and loosened up. It can also transform the means you see or hear things.

What are the unsafe impacts? It can make you really feel exceedingly worried or extremely anxious or make you panic. It can additionally make you feel extremely questionable about everybody (paranoid). It makes you most likely to develop a mental illness such as schizophrenia. Driving intoxicated of cannabis makes you more likely to have a crash. It can make your brain work less well, so you don't concentrate or remember things too. It can be specifically hazardous for people with a heart problem. If you are attempting to have an infant, it can make you less productive.

Just how much does it set you back? The street cost is around £ 5-7 per gram.

Drug

Various other names: fracture, coke, white, toot, pebbles, freebase.

Just how is it taken? The powder is called coke, as well as is typically smelled up the nose.

It makes people really feel super-confident and also

alert. After a big high, there adheres to a 'come-down' or reduced.

People can do harmful things when they feel a lot more certain than they should. It makes the heart beat quicker as well as it can occasionally trigger very high blood stress or heart attacks. The drug can come to be really habit forming, and individuals who utilize it consistently yearn for much more.

Just how much does it set you back? The street cost is around £ 45 per gram for cocaine powder.

Ecstasy

Various other names: E, crystal, dolphins, Superman, pills, Mitsubishis, MDMA, Mandy, brownies.

How is it taken? Euphoria is usually swallowed as a pill.

What does it do? It makes you feel happy and high as well as energetic. It can make colors and also seems more extreme. It can provide you sensations of love as well as love in the direction of individuals around you momentarily. It is commonly utilized by individuals that go clubbing, so they have the power to continue late into the evening. The results last for a number of hours.

The after the result, or 'comedown,' can make you feel very reduced. It can cause stress and anxiety, memory, and also clinical depression problems. In some cases, it can cause issues with your immune system, heart, kidneys or liver.

Just how much does it cost? The street price is around £ 6 per tablet.

Heroin

Various other names: H, smack, skag, gear, brown.

Exactly how is it taken? Heroin is typically liquified right into a liquid and then infused. It can additionally be smoked or grunted.

What does it do? Heroin is made from the opium poppy. In its form for medical usage, it is called diamorphine as well as is made use of as a very strong painkiller. It makes you feel tranquil, satisfied as well as unwinded.

What are the dangerous impacts? Once individuals are addicted, it is extremely hard to quit utilizing it; Heroin is extremely addictive; They can stop breathing and also lose consciousness or die when people overdose on it. Making use of infected needles to infuse heroin can trigger the spread of hepatitis or HIV. It can damage the blood vessels.

How much does it cost? The street cost is around £ 52 per gram.

Ketamine

Other names: K, vitamin K, super K, special K, green, donkey dirt.

Exactly how is it taken? It can be ingested, grunted, or infused.

What does it do? It can make you really feel

extremely loosened up. It is really strong medicine. It can make you really feel as though you are somewhere else rather than in your body. It can give you hallucinations, as well as influence the means you see time and room. This result is called a 'trip.' A journey can be an excellent or bad one.

It can make you puzzled and frightened. It can make you really feel ill. It can harm your bladder as well as make you feel like you need to pee much more frequently as well as urgently.

What does that cost? Prices of ketamine around $20 per gram.

LSD.

Several other names: lysergic acid diethylamide is the chemical term for LSD. It is also commonly known as acid. Certain words include blotter, tripper, light, famous people, rainbows, and paper champignons.

How is it taken? It is swallowed as a tablet or declines of fluid.

What does it do? Like ketamine, LSD triggers bad or excellent trips. A great journey can make you feel loosened up and also satisfied. A bad trip can make you really feel scared and make you panic. It can make you see points that aren't there (hallucinate), and also, these can be good things or bad things.

It can make you confused and also anxious. Individuals can be extra most likely to self-harm when they have a bad journey.

Just how much does it set you back? It sets you back around £ 5 per tablet.

Magic mushrooms.
How are they taken? Magic mushrooms grow wild in the UK. Both kinds are liberty caps as well as fly agaric. They are eaten raw or dried as well as used in beverages.

A great trip makes you feel delighted, cooled out, and also certain. Some people discover magic mushrooms make them extra imaginative or imaginative or sensitive, or they have a sensation of spiritual knowledge.

They can make you really feel sick or offer you tummy ache or diarrhea. You could put on your own in threat when you really feel separated from reality. Just how a lot do they set you back? If purchased, they set you back regarding £ 5 for a handful.

Mephedrone
How is it taken? It can be grunted, or covered in paper and afterward ingested (' bombed').

It is a stimulant with similar impacts on amfetamines. It makes you really feel confident and blissful and talkative.

What are the damaging effects? It can make you really feel woozy or ill or give you headaches. It can make you distressed or questionable. It can cause your heart to race. Occasionally it creates fits. It can damage your throat, nose, or mouth. It can stop you from resting.

How much does it set you back? The price is around £ 10-15 per gram.

Methamphetamines

Various other names: yaba, glass, meth, crank. 4-methylamphetamine is another solid kind of amfetamine, additionally called ket Phet or phet ket.

How are they taken? Methamphetamines can be swallowed, snorted, infused, or smoked. The crystal form is the strongest and most habit-forming.

What do they do? They cause a solid 'high,' which lasts for 4-12 hours (longer than split cocaine), followed by a serious 'comedown.' They may make you feel energetic, as well as awake. They might make you really feel a lot more aroused. They might make you less starving.

What are the risky consequences? You may feel very restless, anxious, nervous, hostile, or doubtful. You were increasing your cardiac rhythm and your blood pressure, and your heart arrest is more likely. You that the limitations and are likely to take hazards that you don't typically take–sex with unfamiliar people as an example. They may cause serious psychological or psychological illness. Bloodthirsty, suicidal thoughts or actions can have an effect. Overdose can lead to death. It's very addictive.

How much do they cost? Cost is extremely variable;

however, they can set you back up to £ 250 per gram. They are used much less in the UK than in various other countries such as the USA.

Unstable substances - glues, aerosols as well as gases.

What substances are abused? Aerosols (such as deodorants, air freshener, hairspray), nail polish or nail gloss eliminator, adhesive, paint cleaner, footwear gloss, gas, cigarette lighter fluid. It can be called sniffing, tooting, huffing, cleaning, or humming gas.

Just how are they taken? The material is inhaled or if in a spray kind, can be sprayed to the rear of the mouth.

They may make you feel fired up and happy. You may really feel high and giggly.

They can offer you trouble sleeping and make you feel tired. They can make your heart quit. Sometimes the gas at the back of your throat can make it swell up so you can not take a breath.

Leisure medicines and also the law

The Misuse of Drugs Act of 1971.

The Misuse of Drugs Act was a regulation that comes in 1971 in the UK to try to stop the use of dangerous medications. It divides drugs right into 3 categories - C, b, or, depending upon how unsafe they are believed to be. Each of the classifications then has different penalties for those found guilty of usage or supply.

Offenses under the Misuse of Drugs Act consist of:
- Possession of medicines.
- Supply of medications.
- Possession with intent to supply one more person
- Offering to provide an additional individual with drugs
- Manufacture of medicines.
- Export or import of medications.
- Allowing your building to be made use of for the usage, supply, or manufacturing of drugs

Class A: medications are one of the most unsafe. They consist of cocaine, heroin, euphoria, LSD, methadone, and magic mushrooms. The maximum charge is 7 years behind bars plus a penalty for possession, and also life imprisonment plus a penalty for supply.

Classe B: medicines include amfetamines (apart from injectable kinds), cannabis, mephedrone, codeine as well as barbiturates. The maximum fine is five years in prison plus a fine for ownership, and also 14 years behind bars plus a fine for supply.

Class C: medications include minor tranquillizers and also anabolic steroids. The optimum charge is 2 years in prison plus a fine for ownership, and also 14 years behind bars plus a fine for supply.

Numerous medicines have clinical or scientific usages, so they are positioned into one of five routines by the Misuse of Drugs Regulations of 2001. These enable some drugs to be lawfully made use of in particular scenarios.

Schedule 1 medications are those who have no legit medical function. These are purely regulated and can just be made use of with a unique Home Office permit. This classification includes cannabis, euphoria, LSD, and raw opium.

Set up 2 medicines that can be recommended by medical professionals and also they are legal to possess if they have actually been recommended. They consist of amfetamines, drug, dihydrocodeine, heroin, methadone, morphine, opium in a medical form, pethidine, and also methylphenidate hydrochloride (Ritalin ®). These have to be cared for in a particular method by pharmacies.

Arrange 3 drugs that can also be prescribed legally. They include teams of medications called barbiturates and benzodiazepines, which make you sleepy. There are special rules leading prescriptions. The bulk of narcotics in Schedule 4 contain tranquilizing agents and anabolic steroids. You are legal if you have a prescription. It is not, however, acceptable to provide them to anyone.

Schedule 5 medications are those who are not likely

to be abused. This consists of specific cough medications and also moderate painkillers.

' Drug driving' - the brand-new offense
This refers to driving, attempting to drive, or supervising of a lorry while having a specified regulated medication in your body, above a defined limit. This came into a result in March 2015. The drugs covered by this offense include cannabis, cocaine, MDMA (ecstasy), LSD, ketamine, methylamphetamine, and heroin.

Roadside medication screening devices utilize spit (saliva) to identify if the person driving or in control of the lorry has taken medicine as listed above. Adhering to a favorable result using saliva, you can then be asked to give a blood sample for telltale functions, to allow prosecution for the new offense (if you are above the defined limit).

How can I inform if my child or my pal is utilizing entertainment medications?

Some indications which MAY show somebody is using drugs consist of:

- They may come to be extra irritable or have state of mind swings (yet remember, teenagers can be moody for lots of factors!).

- They might start staying out later or socializing with new good friends.

- They may seem wearier and also have trouble focusing, or lose interest in things they normally take pleasure in. Schoolchildren might begin doing much less well at school or stop doing their normal hobbies. Adults may overlook their normal obligations.

- Sores around the mouth or nose
- Losing their hunger
- Odd things around the house, such as torn cigarette packages, little sealable plastic bags, silver aluminum foil, needle covers, vacant aerosol
- Needing more money, yet absolutely nothing to reveal for it. Money in the house going missing.
- Regularly entering into difficulty
- Being much more distressed or worried than normal
- Bloodshot eyes; pupils bigger or smaller than normal
- Unusual body scents.
- Shaking, poor co-ordination

Exactly how can I get assistance if I have a problem with leisure drugs?

You are out your own. Lots of individuals have attempted taking medicines and require aid to quit. This may be the beginning of ending up being dependent on it if you locate you are making use of a drug a lot more and also much more often. If you keep desiring you had not afterward, or find you are encountering difficulty in various other areas of your life, there are lots of means to get help.

If that helps you function out just how to quit, assume concerning why you are using the medications and see:

- Is it due to stress from other individuals? If so, you could be able to think about hanging out with various other buddies.
- Is it since it makes you really feel far better regarding issues in your life? You could be able to acquire help by chatting about those troubles with someone if so.
- Is it due to the fact that you are tired? If so, you could look for a new hobby or skill to take up that would be much healthier and delightful.
- Is it because you can't quit since you long for much more? If so, you might need aid from one of the numerous organizations that assist people in giving up.

There are lots of different ways of looking for aid; it depends upon you, which may fit you the best. Often it is handy to speak to the people who care regarding you - your parents, your friends, your teacher, your GP. There are numerous organizations that help people intending to stop taking medicines. You can choose to contact them in numerous ways - online for real-time conversation, or by email, or by phone, or by visiting personally. Or you can just check out the information on their web sites. Several of these, such as 'Frank,' 'Turning Point,' or 'Know the Score,' are listed below. Your GP would be able to recommend the appropriate location to go for help.

Don't be frightened of telling people concerning your medical issue. They are most likely to be eased. You want to do something to quit.

Exactly how can I obtain assistance for one more individual who has trouble with entertainment drugs?

Talk with them initially. Be truthful about your problems, and also talk about the risks of drug-taking and the concerns you have. Attempt to understand why they are taking drugs. Attempt not to be critical. If you understand why it is happening, you are more likely to be able to aid them in quitting.

Either on your own or with the individual you are concerned concerning, you can talk and go to your GP as well as ask advice. Or, you can speak to among the lots of organizations that aid individuals who are misusing drugs. Numerous of these are listed below. They can give you recommendations and also support. College nurses, teachers, or social employees might also be able to aid as well as advise. It is not an unusual problem, so you are not alone. There are experts with lots of experience in helping other individuals in the exact same watercraft.

USAGE OF PSYCHOACTIVE SUBSTANCE IN RELIGION OVER THE YEARS

Amanita muscaria had spiritual importance in ancient India, and travelers videotaped its use as late as the 18th century in Northeastern Siberia. In Central America, psilocybe mushrooms were made use of for the exact same purposes. Indigenous peyos (Lophophora williamsi) were used in pre-Columbian Mexico as were in Southwest Mexico to increase spiritual self-questioning.

A number of religious beliefs have actually used particular types of drugs throughout history as a component of spiritual ceremonies, routines, and customs. For the adherents of these beliefs, medications are believed to bring essential visions and also to aid individuals attach to the spiritual globe or to a higher power. Many faith communities have strong beliefs toward drugs and alcohol as they truly feel that drugs can be associated with a higher power.

Religious convictions using drugs

The use of Peyote is an important part of certain

cases, with 80 chapters in the United States. Members should be advised not to use alcohol.2.

In the 1930s in Jamaica, the Rastafarianism was founded. The supporters worship Haile Selassie as their messianic savior and also think that he will surely return to Ethiopia and take the black area back. Rastafarians, or Rastas, feel that the black neighborhood has actually been subdued by enslavement as well as colonization. Participants are urged to follow a rigorous vegan diet as well as avoid alcohol. However, the ceremonial use of cannabis to enhance spiritual awareness is permitted. One renowned Rasta, Bob Marley, was pivotal in spreading understanding of the religious beliefs through his music throughout the 1970s. The faith has more than 1 million participants around the world.

Hinduism is just one of the globe's oldest faiths and presently has greater than 900 million fans, the majority of whom are from India. Hindus believe in one supreme God with numerous deities related to him. They likewise rely on reincarnation based on fate, or just how a person resided in a previous life. In general, the religion disapproves of controlled substance use, yet cannabis and also a plant-derived intoxicating drink called Soma (not to be perplexed with the modern muscle depressant of the very same name) have historically been made use of in prayer. Fans think that all beings, including pets as well as plants, include a spiritual significance. The hallucinogenic medicine iboga is occasionally made use of in Bwiti spiritual events.

Drugs Used
Peyote is derived from a small, spineless cactus plant which contains the hallucinogenic medicine mescaline. The crown or top of the cactus is cut, dried out, and afterward either eaten, ground right into a powder and also ingested, smoked with marijuana or tobacco, soaked in water, or brewed in tea. The effects of peyote include ecstasy, distorted detects, hallucinations, distortions of room and also time, and also altered body picture. The medicine is utilized by members of the Native American Church in the southwestern United States and also the Huichol Indians in Northwestern Mexico as a component of spiritual events.

Ayahuasca is a hallucinogenic drink originated from Amazonian plants containing DMT, a medication that can trigger hallucinations. Some South American tribal communities use ayahuasca in spiritual routines. A lot more lately, tourists have actually traveled to the area to experiment with the medicine.

Psilocybin, also known as magic mushrooms or shrooms, is stemmed from fresh or dried mushrooms discovered throughout the United States, Mexico, and South America. Users may ingest the drug orally by making the mushrooms in tea or combining them with food to mask their bitter preference. The medicine's impacts can consist of hallucinations, trouble differentiating dreams from the truth, and also anxiety. The medicine is utilized by some native societies throughout Mexico as well as Central America, including the Mazatec Indians of Mexico, as a part of spiritual events.

Marijuana has actually been used throughout the background by some religious groups. Rastafarians refer to the medicine as "knowledge weed" or "divine herb," as well as consider it to have great religious value. Some followers of Hinduism likewise make use of marijuana for spiritual functions.

Salvia divinorum is a herb found in Southern Mexico, Central, and South America. Chew, burn, spray or smoke its leaves or eat the juices of the plant, Salvia may be taken. Salvinorin A may cause hallucinations, distorted sensory experiences, brilliant colors, shapes, motion, and even bright lights as a major ingredient in the medication. The drug has actually been described by the Mazatec medicine men in Oaxaca, Mexico as "leaves of the Virgin Mary" and also made use of for spiritual and also medical objectives.

Kava, or Piper methysticum, is a plant native to the Pacific Islands. Kava essence is made from the part of the plant that grows underneath the ground. The impacts of kava can include modifications in mood, a feeling of health, and also muscle leisure. The medicine can additionally cause hallucinations that might last 1 to 2 hours.3 Some societies in Oceania and also the Pacific Islands to utilize the medicine in spiritual rituals.

Ibogaine is a psychedelic medication derived from the bark of the Tabernanthe iboga plant in Africa. The Babongo people of Africa praise the medication as a source of spiritual expertise and also utilize it as a component of religious ceremonies.

Fly agaric mushrooms are both the first known hallucinogenic mushrooms used by guides and originate from the Amanita muscaria fungus. Tribes in Northeastern Siberia, Mexico, and also Guatemala, and Native Americans in the United States and Canada use the medicine. In Europe, Christians have also used fly agaric mushrooms for spiritual purposes.

Why Are Drugs Used?

The Native American Church sights peyote as a gift from God. During all-night peyote rituals, the medicine is eaten or made in tea as well as passed around from participant to participant.

Rastafarians check out marijuana as spiritual and also utilize the medication during Rastafari reasoning sessions to boost neighborhood in between participants as well as experience relaxing and also spiritual visions. Prior to making use of marijuana, members collectively hope with one an additional.

The Amazonian individuals of South America use ayahuasca to connect with the spiritual globe as well as for knowing as well as recovery functions. Typically, participants are unable to move for 4 to 6 hrs after taking medicine and exist down to pay attention to songs and chanting.

The Mazatec shamans of Oaxaca, Mexico, use salvia Divinorum in religious events and for healing objectives. Salvia is made use of by the Mazatec witch doctors to predict the future and receive magnificent answers pertaining to family members, pals, and enemies. Salvia is also used to manage physical conditions such as migraines. Throughout rituals, the medication is squashed approximately remove its juices and after that eaten or consumed with water.

Marijuana is utilized in Hinduism to promote spiritual experiences. Hindus associate marijuana with the god Shiva, which is believed to have provided the drug to mankind as a sign of appreciation.6 It was very first pointed out in The Vedas, a frightened Hindu message, as far back as 14000 BC. Cannabis was named a sacred plant as well as considered a source of pleasure, liberty, as well as joy and also a means to relieve anxiousness. The medicine is frequently consumed as a beverage or combined with nuts, flavors, milk, or yogurt.

The Bwiti people of Africa utilize iboga throughout

religious ceremonies. Bwiti shamans make use of iboga to see right into the future, talk with animals and plants, attach with the left, and also cure sickness.

Pacific Islanders utilize kava during spiritual routines and events to improve family members' connections, affirm one's rank in society, and connect with spirits.12 The drug has been utilized for over 2,000 years. Kava extract is prepared by softening the roots with water or coconut milk and eating the fluid.

European medicine men and tribes in Siberia utilize fly agaric mushrooms throughout religious events. The medication is illustrated beside the Tree of Life in a fresco in an old French Church, recommending that it has been used since the start of Christianity.

Faiths That Forbid Drug Use

Islam opposes alcohol and medicine usage unless it is medically shown. Tobacco is not particularly prohibited, but its usage is highly inhibited.16.

A Lutheran Catholic church called on its members to discourage drugs, alcohol, and tobacco, and to help reduce drug issues all over the world. According to the church, it is a member's responsibility to "engage in whatever brings our ideas and bodies right into the self-control of Christ, that wishes our goodness, wholesomeness, and joy" and "considering that alcohols, tobacco, and the irresponsible use of medications as well as narcotics are damaging to our bodies, we are to avoid them.".

It is the Mormon Church or the Church of Jesus Christ the Holy Later-Day that alcohol, illicit drugs, cigarettes, coffee, and prescription drugs, as well as abusing prescription drugs are harmful to the body and also clash with the "Word of Wisdom" of the Lord. This calls for a diet plan rich in fruit and vegetables, grains and meat in small quantities. While the church is highly motivated to refrain from illicit drugs, it helps people who have become addicted.

Creator's Witnesses adhere carefully to the principles of the Bible, including that members prevent any kind of "techniques that contaminate our minds and bodies," such as smoking, misusing drugs, or becoming drunk.19 Jehovah's Witnesses believe that modest alcohol usage is not wrong, but that overdrinking is hazardous as well as upsetting to God.20 In some cases, abstaining from alcohol is advised, such as if an individual can not regulate his or her drinking.

SURPRISING MEDICAL USES FOR ILLICIT DRUGS

Some medicines have been made use of as drugs for most of human history. The medical use of opium is explained from the earliest composed documents. It is composed that the attractive Helen of Troy had actually obtained this potion from an Egyptian queen and that she utilized it to deal with the Greek warriors (" currently she cast a drug into the red wine of which they consumed to lull all discomfort as well as temper and also bring lapse of memory of every sadness").

In the 19th century, laudanum was extensively utilized in kids and grownups, for many signs (sleeping disorders, cardiac as well as infectious conditions). In the very early 20th century, encyclopedias in Western countries still specified that persons in excellent mental and also physical health and wellness could use opium without danger of dependence. Griesinger (1817-1868), a German psychiatrist, one of the founders of modern-day psychiatry, suggested the use of opium in the treatment of melancholia.6.

Immoral drugs are bad for you. Nearly every doctor will recommend avoiding recreational substance abuse because it can result in long-term health issues and drug abuse that can ruin individual relationships and also send an individual to an early tomb.

But while drugs can be harmful, illegal medications have been researched for centuries, as well as some have been discovered to have surprising therapeutic advantages.

Whether it's an emotional advantage or treating dependency on one medication with the help of an additional, numerous research studies record the prospective medicinal effects of otherwise illegal drugs.

LSD (Acid)

Lysergic acid diethylamide, called LSD or acid, can be found in tablets, capsules, liquid, or on absorptive paper (visualized left). The hallucinogenic medication creates "trips" that last regarding 12 hours and also include uncertain feelings of panic as well as fright. LSD raises body temperature level, heart rate as well as blood pressure, and has also been connected to "recalls" months after use, it might have a serious effect on alcoholics, according to current research.

The research found people with alcohol problems who took LSD reported greater self-acceptance, awareness and also an inspiration to address their alcohol abuse.

Marijuana

A male smokes a joint at a pro-marijuana "4/20" party before the state capitol April 20, 2010, in Denver. April 20th has become a de facto vacation for marijuana advocates with large events and also "smoke outs" in several components of the United States. Colorado, among 14 states to allow the use of medical marijuana, has experienced a surge in marijuana dispensaries, trade shows and also associated organizations in the last year as cannabis usage has actually come to be extra conventional.

One of the most commonly made use of a controlled substance in the U.S. has found many objectives - recreational, spiritual, as well as medicinal. Uncle Sam even administers free joint containers to four Americans who were grandfathered into an experimental federal government treatment research that checked out cannabis for medicinal factors. Others are just authorized to expand medical marijuana on their own. Marijuana has been found to eliminate persistent discomfort, prevent trauma, as well as have actually even discovered celebrity advocates such as TELEVISION character Montel Williams.

Presently 16 U.S. states and Washington D.C. have clinical cannabis legislation on the books.

Marijuana usage has actually also been connected to

lasting mind troubles, a threat to psychotic signs and symptoms, and harmful vehicle collisions.

MDMA (Ecstasy)

MDMA, referred to as XTC, x or euphoria, is a synthetic drug that produces temporary sensations of emotional heat, physical energy, as well as enhanced sensory perception. However, it can likewise trigger nausea or vomiting, chills, muscle mass cramping and also blurred vision. The medication, which grew in popularity as a club drug, was found to potentially hold the secret to far better therapies for deadly blood cancers cells such as myeloma, lymphoma, as well as leukemia.

One more research found MDMA integrated with therapy might assist deal with trauma (PTSD).

Drug

Fracture, drug, coke, medications, illicit cocaine, recognized in its crystal kind as "crack," is a highly addicting energizer that is snorted, infused, or smoked. It generates sensations of euphoria while raising body temperature, blood pressure, and also heart rate - consequently likewise raising risk for cardiovascular disease, breathing failure, strokes as well as seizures. Premature death can take place on the first use of the drug.

Regardless of its dangers, the drug has a history of medicinal use, when utilized as an energizer for those wasting away from condition or morphine addiction

- the latter was particularly common adhering to the Civil War - as well as believed to be a general magic bullet, offered as restoratives by pharmacologists. Cocaine was additionally among the earliest anesthetics used for surgical treatment — some present anesthetics, such as novocaine, usage safer variations of the chemical without the psychological results.

Psilocybin (Magic mushrooms)

Psilocybin is a hallucinogen located in certain kinds of fungus, commonly referred to as psychedelic or magic mushrooms. While eating these mushrooms produces short-term medication "trips," research has actually found that magic mushrooms might lead to a long-term higher sense of health and help deal with clinical depression.

A further investigation found that medicine could give people a more "friendly" attitude, making them more likely to stop smoking or treat stress, anxiety, and anxiety in patients with cancer cells.

" We're not saying go out there and consume magic mushrooms," Professor David Nutt, a neuropsychopharmacology scientist at Imperial College London and also magic mushroom research writer, said. "But ... this medicine has such a basic effect on the brain that it's reached be purposeful - it's got to be informing us something regarding how the brain functions.".

Ibogaine

iboga plant, ibogaine, hallucinogenIbogaine is a hallucinogenic drug that's discovered in African Iboga shrubs and also is commonly considered spiritual rituals. In spite of it being a powerful hallucinogen that's prohibited in the U.S., some dependency physicians in various other countries are resorting to the plant to treat heroin as well as opioid addiction.

One ibogaine scientist, Dr. Kenneth Alper of NYU Langone Medical Center, told HealthPop last summer season that the strategy is primarily used for people that fall short even more conventional therapies. The medicine is thought to help addicts by helping them through possibly dangerous withdrawal symptoms, as well as reportedly altering drug-seeking habits in some addicts.

Ketamine

Ketamine, additionally called "Special K," is a club-drug that puts customers in a trance-like state called a "K-hole." The medicine is generally used as a pet cat depressant by vets, and also may trigger aesthetic hallucinations, dazzling dreams, confusion, and disorientation, raised drool, as well as troubles with heart rhythm as well as breathing.

Yet recent research of clinically depressed individuals at Ben Taub General Hospital in Houston found the medication treated people with suicidal anxiety throughout the vital stretch when depressed clients were most at risk.

" It was a various experience from really feeling high. This was feeling that something has been gotten rid of," stated Dr. Carlos Zarate, a ketamine scientist at the National Institute of Mental Health.

DEVISING MORE POTENT COMPOUNDS

The fermentation of grains consisting of starch produces a beer with an alcoholic web content of around 5%, whereas the exact same process with grape sugar returns a glass of wine consisting of up to 14% alcohol Distillation made it possible to obtain beverages with much higher alcohol content. People could drink alcohol with a strength of 50% and even more, making it easier to end up being drunk. The construction of stills, connecting an alembic to boil down a liquid with arrangements to condense the vapor created, appears to have started only in the 11th or 12th century around the medical institution of Salerno in Italy.11 Distillation, though it did not produce the troubles with alcohol, can intensify them.12 The "water of life," as it was called in many languages (Latin aqua vitae), overcame Europe at a great rate.

Alcohol without fluid (AWOL) is an extra current procedure that permits people to take in liquor (distilled spirits) without actually consuming fluid. The AWOL maker vaporizes alcohol as well as blends it with oxygen, enabling the customer to take in the mix. Vaporized alcohol enters the bloodstream faster, and also its results are a lot more prompt than its fluid equivalents, creating a blissful high. Traditionally, coca fallen leave is chewed approximately manufacturing in Southern America, for instance, by Andean miners to decrease fatigue. At the other pharmacokinetic extreme, the cigarette smoking of crack drug produces short-lived and intense impacts that are felt nearly instantly after smoking. Opium is another example of a material whose pattern of usage changed in the last centuries, from a drug utilized for pain relief as well as an anesthetic to a material associated with misuse as well as dependency. Opium's capacity to induce dependency was possibly strengthened by the current purification of morphine, and the synthesis of heroin, more powerful compounds that are readily available for the shot. Cigarcttcs, which allow pure nicotine to be quickly taken in the right into the bloodstream and to reach the mind in a couple of secs, were linked with even more dependancy than previous modes of tobacco use (snuff, stogies, eating) which did not advertise deep breathing into the lungs.

THE HISTORICAL ORIGINS OF ADDICTION MEDICINE
Chronological turning points
Aristotle recorded the effects of alcohol withdrawal as well as cautioned that alcohol consumption throughout maternity could be harmful. Calvinist

theologians who used explanations of the condition of compulsive alcohol consumption, which were later adopted by doctors, also bear responsibility for the advancement of addiction medicine in modern times. Dependence has become a major public health problem with the colonial era, the industrial revolution, and international trade.

In the 18th century, opium's addictive possibility was acknowledged when a multitude of Chinese individuals came to be addicted, and the Chinese federal government tried to subdue its sale and usage. In Europe, the functioning classes were threatened by alcoholism. At that time, psychiatry had grown right into a scientific discipline, established nosological categories, and taken depend on social concerns. The 18th-century U.S. psychiatrist Benjamin Rush argued that the loss of self-control, primarily because of beverages and not drinkers, meant that drinking was unregulated. His remarks refer only to high levels of alcohol; in his opinion, a glass of wine and also beer were helpful calming drinkers. In german-speaking nations, Constantin von Brühl-Cramer was the most influential physicist credited with coining the term "dipsomania"[1819]. Devoted clinical papers were published in the 19th century. Devoted clinical journals were produced in the 19th century. The Journal of Inebriety showed up in the United States in 1876, while the British Journal of Addiction was very first released in 1884. Emil Kraepelin, the medical professional who exerted the greatest influence on the shaping of modern psychiatry, combated alcohol with extreme devotion. In the early 1890s, he published the first psychometric data on the effects of tea and alcohol.

Prior to that, he had actually been a moderate enthusiast, identifying alcohol's relaxing and mood-elevating results, as in this letter to the psychiatrist August Forel in December 1891: ". Sigmund Freud, a modern of Kraepelin, laid the ground for the mental approach to addiction. Freud composed in a letter to Fliess in 1897: "it has actually dawned on me that masturbation is the one major behavior, the "primitive" addiction and also that it is just as an alternative and also replacement for it that the other addictions - for alcohol, morphine, tobacco, etc. - come into presence."

An effect of the emotional strategy is that the dependency to various substances (alcohol, narcotics, etc.) as well as also to certain types of behavior, such as gaming, have actually been gathered with each other under a common denominator, and pertained to as different expressions of a solitary underlying syndrome. In the 20th century, addiction medication was actually enriched by (i) analysis categories and (ii) neurobiological and also genetic study., peyote);

euphoriants (cocaine; opium derivatives such as morphine, codeine, heroin); and also hypnotics.

Also, pet study and practical brain imaging researches in humans have actually brought about the existing significant theory that all medicines of abuse share a typical residential or commercial property in applying their addictive as well as enhancing results by (i) acting upon the brain's reward system and also (ii) conditioning the brain by triggering it to translate drug signals as potentially prominent or biologically gratifying stimuli equivalent to food or sex. Hints connected with pure nicotine, drug, or morphine turn on particular cortical and limbic brain areas. This conditioning involves the prefrontal cortex and glutamate systems. In rats, this pattern of activation displays similarities to that elicited by conditioning to an all-natural reward-highly tasty food such as chocolate.

Challenged by signs that work as medicine suggestions, the specific experiences desire, as well as the level of volunteer control that he or she is able to exert, may be impaired. This theory is partly originated from Pavlov's conditioning paradigm, where food is related to drug, the animal's drool to cocaine yearning, and also the bell to the medication hint.22 Family, fostering, and twin researches have actually demonstrated the intervention of hereditary factors in addiction,23 significantly in alcohol abuse and also dependency. Hereditary variables engage in an intricate means with the environment.24-26

Dependency - history of a word
The meaning of dependency has actually advanced gradually. Today, addiction is defined by the particular features that are shared by a variety important:
(1) the pattern of administration can progress from use to abuse, to dependancy and also
(ii), as discussed in the previous paragraph, a typical attribute of numerous substances is that they induce satisfaction by triggering a mesolimbic dopaminergic benefit system, and also dependence by devices including adjustment of prefrontal glutamatergic innervation to the center accumbens.

The term "addiction," in its existing clinical meaning, was made use of initially in English-speaking nations, and afterward passed on to other languages that had utilized various other terms previously. Addiction has displaced the words toxicomanie or assuétude in French. Interestingly, words assuétude (from the Latin assuetudo [behavior] had actually originally been presented right into French in 1885 to convert the English dependency.

Non-Latin roots in German usages, such as dependence (reliance), addiction, and intoxication
In Roman regulation and also in the middle Ages,

addiction was the sentence articulated against an insolvent borrower who was offered over to a master to settle his debts with his job.

Therefore, the addict was a person oppressed as a result of debts. According to the Oxford English Dictionary, the term "addict" in the significance of "attached by one's own inclination, self-addicted to a method; committed, provided, inclined to" has been utilized since the first part of the 16th century. Nevertheless, dependency, in its present clinical significance of "state of being addicted to a medication; an obsession as well as require to proceed taking a medicine as a result of taking it in the past" has actually been in extensive usage just because the 20th century In clinical English, dependency replaced older terms, such as "inebriety.".

The difference between the terms reliance as well as addiction has long been debated. The meaning of these terms among public health and wellness professionals can just be understood in the light of their historical advancement. Addiction is defined as "strong dependence, both physiologic and psychological" in Campbell's psychiatric dictionary28 In 1964, the World Health Organization recommended that the term substance abuse change addiction as well as adaptation because these terms had actually failed to supply a definition that could put on the entire variety of drugs in use. Historically, the archetypal design of dependency was opiates (opium, heroin), which induce clear tolerance (the requirement to boost doses), extreme physical withdrawal signs and symptoms when use is ceased, and also have major consequences for the social, specialist, and also the domestic performance of users.

The spread of the principle of addiction to various other substances, notably nicotine, occurred only in current decades. The medical diagnosis of tobacco reliance or dependency did not exist in the Diagnostic and also Statistical Manual of Mental Disorders, 2nd ed (DSM-II, American Psychiatric Association in 1968).

(DSM-III-R), committee participants differed as to whether "dependency" or "reliance" should be taken on. A ballot was taken at a board meeting, as well as the word "dependence" won over "addiction" by a solitary ballot! As directed out by O'Brien, the term "dependency" can define the uncontrollable drug-taking condition as well as distinguish it from "physical" dependence, which is normal and also can happen in any person taking medicines that influence the brain.

UNDERSTANDING PSYCHOACTIVE SUBSTANCE ABUSE AND VIOLENCE

Drug and alcohol abuse has actually made complex effects on the human mind and also human habits; as far back as 1995, the Journal of Health Care for the Poor and Underserved described the partnership between drug abuse and violence as a case of "reason as well as effect." The connection between medication alcohol addiction, dependency, and physical violence crosses several thresholds (private psychology, public wellness, and domestic violence, among others), as well as is critically important in recognizing the extent of exactly how controlled substances can affect people.

Medicines and Violent Behavior
The Journal of Substance Abuse Treatment noted that even more than 75 percent of individuals that start therapy for medicine addiction report having actually executed various acts of physical violence, including (but not restricted to) mugging, physical attack, and also utilizing a tool to strike an additional individual. Examining sex differences, the Journal of

Studies on Alcohol and also Drugs discovered that prior to looking for therapy for substance misuse, the price of violent acts was as high as 72 percent among males and also 50 percent among ladies.

Why is there a link between alcohol misuse and also violence? Testing individuals that reported that they were prone to hiding their mad sensations, scientists observed a 5 percent boost in terrible actions that complied with a 10 percent rise in drinking to the point of obtaining drunk.

The scientists kept in mind that "only a tiny portion of all alcohol consumption events include violence," but the possibility of being violent while drinking appeared to be based on just how well individuals who consume alcohol can handle their temper while they're sober. Considering that drinking alcohol can lower inhibitions, urge dangerous habits, and also rob individuals of their self-constraint, a private with unreleased craze can act out when completely intoxicated.

Medicine Addiction and Suicide

The compulsion toward high-risk habits represents the connection between alcohol abuse and the threat of self-destruction, which additionally overlaps with the propensity for physical violence. Alcohol addiction, Clinical, and also Experimental Research kept in mind a perception among individuals that look for therapy for drug abuse issues that the lack of ability to control their violent actions was gotten in touch with a boosted chance of a past self-destruction attempt. Researchers writing in that journal thought that people that battle to manage their rage are most likely to act on impulse and also may, therefore, be much more violent to themselves than to others.

To that factor, the World Health Organization specifies self-destruction as a type of "self-directed violence," and also studies on the topic of exactly how chemical abuse affects physical violence have found a solid link between addiction as well as "self-directed physical violence." The European Archives of Psychiatry and also Clinical Neuroscience journal wrote of exactly how individuals with histories of alcoholism as well as past hostile actions "are more likely to report suicidal thoughts or past self-destruction attempts." People who had attempted to devote suicide had a tendency to report higher instances of depressive problems, which they might have tried to deal with by abusing alcohol or medications as a form of self-medication, thus strengthening the spiral.

In 2010, scientists writing in the Addictive Behaviors journal noted that people who had a psychological dependency on alcohol, and likewise had a background of self-destruction attempts, revealed "greater hostile and impulsive actions." In research of over 6,000 individuals registered in therapy for their addiction, individuals that had actually dedicated acts of severe violence (such as rape, murder, or attack that caused the severe injury) were more than 50 percent most likely to report numerous self-destruction attempts. When demographics, mental health, as well as previous instances of victimization were taken a right into account, the fact was untouched also.

Alcohol And Drug Dependence simply kept in mind the "big empirical literature" that attaches alcohol abuse to self-destruction, while additionally discussing that research study has actually determined a web link exists between opioid misuse, intravenous substance abuse, as well as self-directed violence.

Self-destruction Attempts: According to the Archives of General Psychiatry, individuals that have a medical diagnosis of abuse or dependency on drugs, alcohol, or both are 6 times most likely to have actually attempted suicide when contrasted to those who have no such diagnosis.

Residential Violence as well as Substance Abuse

Of all the kinds of physical violence influenced by drug addiction or alcohol addiction, physical violence towards a cohabitant might be among one of the most concerning as well as severe. The U.S. Department of Justice discusses that the abuse in a residential setup is not restricted to physical acts, such as striking, boxing, slapping or drawing hair; "domestic physical violence," as a lawful term, can likewise cover sexual assault (rape, marital rape, dealing with a partner in a sexually violent and also demeaning way, and also molestation), emotional abuse (intentional and also harmful assaults on a companion's self-worth), and emotional misuse (managing the partner, blackmail, intimidating harm to youngsters, physical violence toward animals, as well as scare tactics).

As described by the American Society of Addiction Medicine, violence against an intimate companion can additionally include stalking, social isolation (not letting an intimate partner leave the house, for instance), and denying resources and necessities. Different studies have actually identified substance abuse as a factor in 40-60 percent of incidents of domestic violence, either in precipitating the misuse or worsening it.

The American Journal of Public Health kept in mind that material abuse has a tendency to be extra common amongst females who suffer domestic misuse, even among expecting females who were victims of physical violence. Such is the level of the impact of drugs and also alcohol on physical violence that, on days when misuse happens, physical violence was 11 times a lot more likely to take location.

More than 20 percent of male perpetrators of intimate

companion abuse use alcohol or drugs promptly before one of the most recent and severe events of physical violence. Several research studies have discovered a solid link in between too much alcohol use and also the activity of companion physical violence, although there is not yet an agreement on the cause and effect-- whether the drinking triggers such males to be violent or whether the alcoholic abuse is utilized as a way to reason (or justify) the physical violence.

Unsurprisingly, the targets of intimate partner violence often tend to be primarily women-- 85 percent, according to the United States Bureau of Justice Statistics. Compared to guys, ladies have a 5-8 times higher chance of being abused by an intimate partner. The majority of partner misuse takes place in the house.

Lesbian, gay, bisexual, and also transgender people have greater prices of compound abuse than the general population, commonly due to the prejudice that numerous of them encounter in their lives, as well as this might show up in physical violence lugged out within their domestic unions. Among Native American neighborhoods (where prices of material abuse are "worse than we thought," according to the National Institute of Drug Abuse), practically 50 percent of women experience intimate companion physical violence in their lifetimes.

Unhealthy Coping Mechanisms

In 2002, the Department of Justice reported that most of the targets harmed by their cohabitant did not look for specialist clinical treatment for their injuries, either as a result of pity, because they did not want their abuser to obtain arrested, or because they did not want the drug and alcohol misuse to come to light.

As an outcome of the physical violence, the victims of residential abuse are at a greater threat for enduring psychological health and wellness troubles. Compared to people who do not experience residential violence, victims are 70 percent a lot more likely to abuse alcohol.

Contrary to common belief, it is not only the criminal of the physical violence who is the substance-abusing partner; in many cases, both participants of the partnership participate in an alcohol or drug abuse. Occasionally, just the target of physical violence is a compound abuser (maybe driven to abusing drugs and alcohol as a way of dealing with the physical, mental, as well as psychological trauma of the physical violence).

The confluence of domestic violence and also substance abuse develops a really dangerous scenario for the sufferer in the partnership, for a number of major reasons. The presence of drugs and alcohol in the target's system may make standing up to the opponent unfeasible-- as an example, if the victim is drugged to decrease restraints as well as resistance to sex-related developments intentionally. A target might not have the ability to totally examine the danger being presented as well as may be forced to stay in a harmful situation because of the effects of medications or alcohol.

Sufferers might be unwilling to seek help or report the strikes for anxiety that the drug abuse will certainly bring about apprehensions, pity, or allegations of unreliability. For example, researches reveal that in a number of situations, the sufferers of domestic violence are compound abusers themselves. While the chemical abuse might be a direct result of being revealed to the violence, the situation can likewise put off sufferers from calling the cops.

Substance Abuse and Intimate Partner Violence

The National Criminal Justice Reference Service reveals that in one study out of Memphis, Tennessee, 42 percent of 72 sufferers of domestic attack confessed to utilizing medicines and alcohol on the day of their attack, with 15 percent alone utilizing the drug. As much as 92 percent of the opponents in the report were medication and alcohol users, and also two-thirds of them eaten alcohol and also cocaine together.

In 1996, the Annals of Emergency Medicine

examined 134 murders of females in New Mexico. The crimes happened between 1990 and 1993, and in 62 percent of the murders, the male domestic/intimate companion was the perpetrator. Among the sufferers, 33 percent were legally drunk when they died; nearly 25 percent had medicines in their system at the time of their fatality.

Impacts on Children
The Livestrong Foundation checked out exactly how physical violence driven by substance abuse influences the rest of the family unit, discussing that the factors that link physical violence toward a member of the family, as well as drug dependency, do not always come to an end when the abuser stops the medication consumption. Trauma and lowered happiness can last for several years after the last instance of physical violence or medication usage, with the partnership between the family members as well as each person's specific emotional health and wellbeing completely harmed as an outcome of their abuser's habits.

According to the National Coalition against Domestic Violence, kids who grow up with drug-addicted parents have a greater likelihood of experiencing physical, sex-related, or psychological abuse (all types of residential violence) than kids who grow up in steady residences. Those kids who experience physical violence at the hands of a parent or guardian (either firsthand or by seeing or listening to a drug-using parent misuse one more participant of the family members or a beloved pet dog) have a high chance of establishing their own alcohol and medication problems when they reach the adult years.

Growing up in an Abusive Home
Children have actually been called "the quiet sufferers" of domestic physical violence due to the fact that they are typically unable to appropriately recognize (as well as for that reason speak about) what they see, hear, as well as experience through abusive moms and dad. In addition, the younger a child, the much more prone and also defenseless the kid's mind, so the anxiousness and also stress problem that can develop in the aftermath of a fierce occasion can be long-lasting.

The Center for Nonviolence and Social Justice creates that children who see brother or sisters or moms and dads being terrible towards one another, or who are subjected to this physical violence firsthand, mature with brains that develop in a different way than children that mature in houses where there is no such violence.

A study released in the Neuropsychopharmacology journal (as well as reported on in TIME magazine) showed that teens that experienced residential physical violence and also other injuries during their childhood grew up with problems in specific areas of their brains-- particularly, the regions of the mind that connect feelings to ideas which regulate habits. Consequently, the kids grow up into young adults, and afterward, grownups, who struggle with clinical depression and trauma, have a problem maintaining healthy partnerships, and eventually proceed the cycle of alcohol and drug abuse as an expression of the impulsive and also dangerous behavior that develops as a result of the mental problems.

Among the outcomes of this is that as many as 20 percents of the youngster, as well as teen arrestees in state juvenile justice systems, were intoxicated or high when they dedicated their crimes. Some were apprehended for committing a crime pertaining to drug or alcohol intake or procurement.

Fleeing from Violence

The Drug and also Alcohol Dependence journal described that when children, as well as adolescents, run away from home, they usually do so for a number of reasons; chief amongst those factors is abuse by family members, which creates a useless (and possibly terrible) house environment. As numerous as 80 percent of the homeless young people in America, aged 12-21, usage medicines or alcohol to endure and also cope on the streets. The substance misuse might be voluntary, or it might be forced upon them by gangs and also human traffickers who see the baffled, lonesome youths as a medium for sexual exploitation, and also who utilize medicines as a kind of threat and also repayment.

According to the National Runaway Safe line, as many as 70 percents of teenagers left their home with no planning or prep work, normally since they had reached a point since the abuse to which they were subjected (whether physical, emotional, or sexual) had become unbearable, as well as leaving house was a better threat to remaining. The Journal of Drug Issues researched chemical abuse among adolescents who were runaways, and also scientists ended that teenagers that suffered high levels of physical violence from their guardians or parents had a higher opportunity of being dependent on drugs and alcohol when they left the residence. The shock of getting on their very own, subjected to the elements without convenience or sanctuary, and also still nursing the mental as well as physical injuries of the violence they received contributed to engaging the substance abuse.

The globe that teenagers get in is harmful among rampant medication trafficking (and human trafficking), where drug and alcohol intake is a way of living. In the past, road medicines like heroin or drug may have been the poison of selection, yet in a period where the prescription drug is an extremely valued commodity on the black market, drugs like OxyContin as well as Vicodin are the new items. The potent painkillers numb physical discomfort as well as cause such solid states of tranquility as well as drowsiness that many individuals experiencing tension or injury shed themselves in the numbing daze. Practically a quarter of the youths displaced of their houses in Los Angeles in 2011 mistreated prescription medication.

Socioeconomic Factors

According to the Aggression as well as Violent Behavior journal, "a lot of alcohol and substance abuse takes place amongst persons who are not fierce," yet alcohol and drugs tend to be located in individuals on both sides of a fierce occasion (whether sufferer or aggressor). The researchers writing in the journal explain that there is far more to violent behavior than simply alcohol and drugs; there prevail socioeconomic factors to take into consideration (such as the systemic violence of medicine circulation networks, or the financial uncontrollable violence of making use of force to get drugs or the cash to purchase medications), the setting as well as setting in which people get as well as use drugs, as well as the special organic as well as psychological procedures that drive every facet of human behavior and interpersonal interactions.

Laboratory and also research studies recommend that alcohol has a causal duty to play in fierce actions, yet the level of that duty is considerably varied. The very same relates to stimulants like cocaine as well as amphetamines. That being stated, the relationship between alcohol, stimulant medicines, and fierce actions is "exceedingly complicated" and determined by lots of conditions (a few of them even contradictory) in real life.

A few of the socioeconomic factors consist of crime. The National Council on Alcoholism and Drug Dependence points out that because alcohol use is both lawful as well as pervasive, the material "plays an especially solid function in the connection to the criminal offense." The Department of Justice estimates that alcohol is involved in virtually 40 percent of violent criminal activities, and also of the 2 million convicted offenders currently offering time for their criminal activities, as numerous as 37 percent admit that they were drunk when they were detained. The Bureau of Justice Statistics noted in 1998 that roughly 3 million fierce criminal offenses take place every year where the transgressors were consuming at the time of the case. When the target or the criminal (or both) was drinking, various other statistics reveal that half of all murders and assaults take area.

More than any controlled substance, alcohol was located to be "carefully linked" with fierce crimes, such as murder, assault, rape, as well as abuse of (marital) partner and also children. When the target, as well as the opponent, is accustomed to one an additional, alcohol likewise often tends to be an aspect of physical violence. As high as 66 percent of sufferers who were assaulted by an intimate partner (a term that consists of an existing or former sweetheart, spouse, or sweetheart) told authorities and emergency services that alcohol was taken in prior to or throughout the attack. By contrast, just 31 percent of fierce assaults entailing alcohol was performed by complete strangers. Figures show that nearly 500,000 situations of physical violence between the intimate companions of a partnership involve opponents who had actually been consuming prior to the misuse started; 118,000 episodes of family physical violence (not counting spouses) and 744,000 such instances with colleagues involved alcohol.

Drug Addiction as well as Violent Crime
In speaking about the complicated partnership that exists between medicines and also crime, the National Council on Alcoholism and Drug Dependence expands on the point made by the Aggression as well as Violent Behavior journal; "criminal offense" can cover may measurements, such as use-related criminal offense, economic-related criminal offense, and also systems-related criminal activity.

Use-Related Crimes: use-related crimes are the outcomes of what takes place when people that take in medications act strongly and unexpectedly because of the behavioral as well as mental results of the medicine. The Clinics journal noted that "it is typically approved that crack drug use is related to increased degrees of physical violence," including that there is proof that medication trafficking is positively correlated to boosts in violence.

Economic-Related Crimes: Economic-related criminal activity refers to criminal activities performed by individuals to money their medication routines, such as hooking (which some have stated need to be "upgraded" to being legitimately thought about a form of a terrible criminal offense) and theft. The theft itself covers break-in, store burglary, in addition to robbery and dealing with stolen products-- criminal activities that become part of "feeding the habit." In 2004, 17 percent of state detainees and 18 percent of federal detainees were behind bars because the infraction for which they were jailed was their method of securing sufficient cash to continue their medicine consumption.

System Related Crimes: Crimes that come out of the systems of the drug system include a system-related crime. Several of the crimes under this category do not constitute "solid" criminal offenses by nature, such as the transportion, manufacturing, or even the manufacture or selling of medicines. In Mexico, as many as 120,000 people were murdered in the stalemate between rival gangs and the governments of the United States and Mexico in 2006.

Driving while Impaired

Perhaps one of the most famous forms of medicinal and alcoholic violence drives intoxicated people, the third most commonly reported crime in the US. Yearly, over 1 million individuals are detained for supporting the wheel while impaired; driving intoxicated is the number one reason for death, injury, and also disability for individuals aged 21 and under. Nearly 30% of web traffic deaths are due, according to Centers for Disease Control, to the fact that one or more motorists were drunk at the time, and the National Health Institute states that in the mid 1970's it was up to 60%.

A 2010 survey revealed that one in eight highway school senior citizens admitted to driving after smoking marijuana, the National Highway Traffic Safety Administration reported in 2007 that 8 drivers had been positive for having illegal drugs in their system while behind the wheel.

The increasing appeal of marijuana, lax enforcement criteria, and also entertainment legitimacy in some territories have pressed state and local transport departments to find up with various means to ensure those vehicle drivers that are too expensive do not position a risk to other vehicle drivers, even as some research has shown that states with some action of marijuana guideline have lower prices of web traffic fatalities.

Cannabis is not the only drug that some chauffeurs are consuming before driving. Numerous individuals that are on training courses of opioid or benzodiazepine medicine (recommended for anxiety, persistent pain, or insomnia) make the error of driving while really feeling the sedative results of the medicines. Authorities are stymied, since prescription medicines affect differents in various ways, so no agreement is reached on how much medicines a person can have before driving.

In addition to this, prescription medicine is an important term, which involves drugs that are awareness-raising as well as response time, as well as medications that bring risks to both pedestrians and drivers.

Drug-Induced Violence on College Campuses

Across Universities, substance abuse is a major problem, but alcohol remains the number one poison for students aged 18-24. Greater than 600,000 students annually are assaulted by an intoxicated pupil, and also 95 percent of all the fierce crime that occurs on college building entails either the assaulter, the victim, or both being drunk at the time. 90 percent of all the cases of rape and/or sexual attack on university campuses involve all relevant events being intoxicated at the time of the attack.

The National Institute of Justice warns that "alcohol usage increases the threat of sexual offense," creating that alcohol use by a potential assailant can cause enhanced aggression, aggressive actions, and, most importantly, "a lack of ability to analyze one more person's sex-related interest precisely." Several sufferers of sexual offense report that their drunkenness made them likely to make threats that they would or else prevent; several females likewise reported that their drunk state lowered their ability to place resistance to the sexual developments and, ultimately, the attack executed by the assailant.

For this reason, many universities have actually redefined as well as refocused their alcohol consumption as well as serious crime laws, tipping up enforcement as well as punishment of alcohol-related offenses that straight intimidate the well-being of their pupils.

SUBSTANCE ABUSE TREATMENT OVER THE YEARS

The face of dependency, the assumption of those battling dependency, as well as the treatment of medicine addiction has changed a whole lot throughout the years in the USA. The most current national survey, which was completed in 2013, released that 24.6 million Americans (over the age of 11) were taken into consideration present illegal medication abusers as well as had used an immoral medication in the month prior to the survey, and also 21.6 million individuals were taken into consideration to have a drug abuse or dependence problem in the previous year.

Those addicted to medicines have actually been damned, outlawed, or based on lots of suspicious "treatments" throughout the years, as the public and also the medical principle of addiction has actually been instead liquid. The existing definition of dependency postulates that it is not a failing of ethical personality, but instead a condition of the brain that affects the incentive system, determination, and emotional guideline of an individual. Addiction calls for customized therapy for recuperation and also to stay clear of episodes of regression.

One of over 14,500 addiction treatment facilities in the United States can provide treatment. Throughout America, dependency and drug therapy have been used a long way and grow through brand-new studies and medical findings throughout the years.

Increase of Addiction in the United States as well as the Need for Treatment

Considering that opioid medicines are highly habit-forming, this might have provided rise to the spread of drug addiction in the United States, adhering to the battle. Medication dependency may have been reasonably unusual and also perhaps perpetuated in Chinese opium dens.

By the end of the 19th century, soldiers of the civil war had been made up of some opioid-sufficient individuals, but at the moment, large-8 columns of medium-12 were very likely top and middle class, which could have been prescribed medication for menstrual problems in the first place. Clinical addiction to people considered ethically unstable was no longer restricted.

Individuals suffering from alcohol addiction or medication addiction may have been constrained to an inebriate asylum for an amount of time to help them "dry," as treatment may have concentrated mainly on detoxification, withdrawal, as well as physical stabilization. The New York State Inebriate

Asylum, built-in 1858, really may have been just one of the very first organizations to attempt to deal with alcoholism as an illness, catering to the upper-crust society of New York in the late 19th century till such idea was steered clear of.

The government-run treatment facility was forced to shut and also, later on, reopen as a psychiatric medical facility for the "chronic crazy" where it employed abhorrent therapies like prefrontal lobotomies, hydrotherapy, and electroshock therapy. Inebriate insane asylums and also asylums maintained clients onsite, most likely even versus their will, for a range of different dependency therapy approaches that, in today's times, could also be viewed as barbaric.

Methods Used for Treating Drug Addiction throughout the years the 1800s:

Addiction might have mostly been associated with alcohol or opium; these substances might have been changed with morphine, cocaine, or other intended "medicines" throughout addiction therapy.

1879: The Keeley Cure, or the "Gold Cure," was introduced. This involved injecting remedies having gold, strychnine, and also alcohol into those battling alcohol, narcotics, or pure nicotine addictions. By the end of the 19th century, there mored than 200 Keeley Institutes providing this treatment. The injections might have been pointless; these institutes may have built the structure for team therapy as well as neighborhood support companies.

- 1800-1900's: The use of warm or cold water to "shock" the system with hydrotherapy might have been used to deal with addiction to alcohol; it was frequently utilized to treat mental disorders.
- 1900's: Addiction might have been connected to seasonal depression, or wintertime depression, where individuals may have been dispirited by the cold, dark weather of winter months that was thought to potentially create dependency and also might have been treated with warmth lights or light boxes.
- 1900's: Bromide-sleep treatment was introduced to deal with addiction where people were put into a bromide-induced coma to get up "healed" from addiction. This approach likely had a high fatality price.

- 1909's: In the case of the lethal belladonna plant which could have caused hallucination efforts have been made to "clean" drug addiction as well as alcohol addiction.

- 1899-1903: Antibodies to alcohol, stemmed from equine's blood, were introduced by means of cuts in the skin to those addicted to alcohol. Equine was presented, and quickly thrown out, as a prospective vaccination for alcohol addiction.

- 1907: Several states restricted marriage for people battling addiction. Sanitation may have even been performed to maintain those addicted to substances from continuing as well as procreating dependency right into the future generation.

- 1927: Individuals addicted to medicines might have been intentionally provided a big dose of insulin that would certainly put them into a coma by raising their blood sugar to precariously high levels with the hopes that they would certainly awaken "cured" of addiction.

- 1930's- 1950s: Criminal culprits who struggled with addiction as well as were housed in the Colorado State Penitentiary were contaminated with blisters on their abdominal areas that were after that drained. The resulting fluids were re-injected into their arms as a "treatment" for dependency.

- 1935: Medical withdrawal may have been handled with codeine or subcutaneous morphine shots.

- 1935: To order to treat people not like anything linked to drug, alcohol addiction avoidance therapy was started in the Shadel Sanitary. This therapy resulted in unnecessary stimuli any time there was alcohol, likely by nausea generation as well as by alcohol poisoning.

- 1940-1950: Addiction was thought about to be caused by a disorder in the endocrine system, and also individuals dealing with addiction were infused with adrenocorticotropic essences from the adrenal gland.

- 1948-1952: Tenderness was related to the prefrontal cortex and patterns of behavior. Front lobotomies have been performed on people addicted to drugs or alcohol, or the surgical removal of the frontal lobe.

- The mid-1950s: Electroshock treatment (ECT), where people were over and over again shocked with cords connected to their heads and also bodies, was performed on individuals battling addiction.

- 1950-1960: LSD, the hallucinogenic medication, was utilized to treat individuals experiencing alcoholism.

- Existing day: Even today, the Web generates a huge selection of aversive as well as weird methods and also "treatments" for dependency that can not just make people ill, yet are also mostly inefficient.

Early Criminalization of Addiction and also Negative Effects on Treatment

Society as a whole might have looked the other way and felt that considering that a huge majority of those addicted to these numbing medicines were upper-class white ladies, as well as a result were not a threat to society, their medicine addiction might have been mostly endured. Physicians were also beginning to understand the potential risks of medicine abuse and also addiction, as well as adjustment in the population of people addicted to drugs, might have compelled the hand of the federal government to enact legislation regulating the prescription, sale, and misuse of narcotics.

Public understanding started to transform as drugs

increasingly found their way into city, poor, as well as minority populaces. Culture perpetuated the suggestion that medications were the cause of several criminal acts, consisting of rape, devoted by this demographic as well as cited drug abuse as one of the major reasons. In concern for the safety and security of kids as well as ladies, as well as the expanding residential medicine as well as controlled substance trouble, political leaders may have taken notification. The law on import, selling, and distribution of medicines was adopted in 1914. The law on drugs was introduced. Doctors could no longer prescribe opioids for maintenance purposes, and people addicted to those narcotics could be forced to kill themselves or commit criminal acts in order to try and illegally procure such medicines. Medical professionals were additionally detained for prescribing opioids if they were not regarded medically needed, and also medical professionals were no more able to deal with those addicted to opioids with maintenance dosages out of their offices straight.

Restriction in the 1920s sought to eliminate alcohol as well as mind-altering substances from society generally, although this was located to be inefficient, as well as the laws were repealed by the very early 1930s. Throughout this time around period, community facilities that had been the go-to for individuals fighting opioid or narcotic dependency were closed down. "Ambulatory" opioid dependency therapy, in addition to the brand-new specialty of addiction scientific research, was all but wiped out for a number of years, and also lots of struggling with addiction ended up in prison as opposed to obtaining the assistance they needed.

Change to Medical as well as Supportive Treatment

Around 1924 and 1935 those fighting drug dependence may not have had a great deal of money if they were not from the high classes of society and the brand-new detoxification methods provided by private medical facilities. In 1929 the Porter Act, which demanded that two "narcotic ranches" be created, was passed amid a significant federal prison shortage and no clear solutions to addiction treatment. Program of Public Health.

In 1935, a new prison-hospital with dependence treatment opened in Lexington, Kentucky, while the second was opened in 1938 in Forth Worth, Texas. Until the late 1950s, the two "farms" had been supplying most of the U.S. dependency therapy services. They used a three-pronged strategy, consisting of withdrawal, retrieval, and then rehab, all continued by a clinical as well as mental wellness team of experts. [33] Therapy for dependency moved out of the community-based and "a good reputation" kind facilities to an extra clinical setting.

Consequently, dependency treatment solutions began to move to an extra clinical approach.

In 1935, the Oxford Group was probably the beginning of 12-stage support programs such as Alcoholics Anonymous (AA), a religious movement that valued socially enhanced methods for themselves and communicating them within the group. The Minnesota Model was taken on by the not-for-profit Hazelden Foundation as well as employed a private therapy strategy that checked out dependency as a treatable disease with education and learning, household participation, a 28-day residential keep, as well as continuing assistance through AA engagement.

Effects of Legislation and also Laws on Drug Treatment

The ownership and sale of narcotics were more criminalized in 1952 and 1956 with the flow of the Boggs Act and the Narcotic Control Act specifically, which featured high penalties for drug belongings as well as the sale of narcotics. Youngsters addicted to opioids, and specifically, heroin, became increasingly more widespread, specifically in New York City, in the 1950s, as well as fueled the demand for a teen as well as juvenile drug treatment programs in addition to the principle that addiction was indeed an illness. In 1952, New York City opened up the Riverside Hospital specifically for teens addicted to drugs, although the programs proved greatly inadequate, and also, the facility was not also open for 10 years. In 1958 Synanon in California was the first of the first options for long term properties, provided that the repetitive rates and healthcare areas (TCs) were born.

TC's were, and still are today, domestic communities where individuals dealing with medicine dependency stayed for an extended period of time with groups of people with like situations. They are self-supporting areas that preserve abstinence with self-help and also encouraging techniques. When they first appeared, TCs did not allow for any kind of sort of mind-altering drugs, much in the blood vessel of AA technique; nonetheless, today, TCs might permit making use of upkeep drugs when necessary.

In the 1960's methadone was viewed as a replacement drug for opioid addiction, because it was a lengthy opioid that could be substituted by shorter ones, for example heroin. In 1964, the Narcotics Addiction Rehabilitation Act (NARA) of 1966 provided neighborhood and also state federal governments with government help for drug therapy programs planned for those addicted to narcotics.

While it was still known to be beneficial for drug addiction, methadone also saw a high price for abuse and diversion. It was considered effective. Then in the 1970s, with a Special Action Office on Drug Abuse Prevention (SAODAP), President Nixon controlled the dispensation of the opioid antagonist and placed it under federal control over his War on Drugs. The Comprehensive Alcohol Abuse as well as Alcohol Prevention, Treatment and Rehabilitation Act of 1970 helped to improve drug recovery by recognizing it as a possible condition instead of an ethical stop in personality function.

Protection of Drug Addiction Treatment and also Effects on Services

Between 1964 and also 1975, the insurer began to acknowledge dependency as a treatable condition and began providing coverage for therapy for those battling dependency. By the 1980s, medicine dependency therapy, as well as alcoholism treatment, were ultimately seen as similar, as well as treatment initiatives were merged.

Special services for the population of the elderly, gays, women and youth, and those dealing with coexisting mental health and well-being started to be established regularly in 1985.

In 1987, despite President Regan's renewed War on Drugs project that looked for to punish drug abusers, the American Medical Association (AMA) proclaimed drug dependence as a legitimate illness as well as demanded that it be treated no in a different way than other medical disorders. Miami became residence to the very first "medication court," assisting nonviolent culprits to get involved in dependency treatment programs as well as stay clear of prison time.

Hospital-based inpatient therapy facilities were forced to shut their doors in between 1989 and 1994 after insurance stopped paying benefits. Addiction solutions were rolled right into behavioral health services in addition to mental health and wellness and also psychiatric conditions, unlocking to a more outpatient or intensive outpatient strategy in contrast to mostly residential therapy.

The introduction and also revival of cocaine as a significant medicine of abuse with high rates of addiction might additionally have actually triggered more extensive outpatient treatment options. By the mid-1990s, public funding, as well as Medicaid receivers, were offered with extensive outpatient services as the recommended approach of treatment.

The Drug Addiction Treatment Act (DATA) of 2000 enables the office-based treatment of opioid and also narcotic dependency with clinical upkeep medications, and also the prescription of illegal drugs developed to help with detox and also prevent relapse, such as buprenorphine opioid agonist products. It likewise permits even more freedom in treating dependency by a recognized facility.

Perhaps the biggest piece of healthcare reform

pertaining to psychological health and addiction treatment came using the Affordable Care Act (ACA). As of 2014, the ACA includes chemical abuse and reliance as one of the 10 important health advantages to be dealt with as a clinical problem and also covered by insurance policy similarly as various other or surgical clinical procedures.

Modern Drug Rehab

Scientific research study has been ongoing for many years into the causes, therapy, as well as optimum recuperation efforts for substance abuse and addiction. The clinical community today mainly advertises the condition concept of dependency-- that mind chemistry is changed via routine drug abuse, resulting in behavior modifications and also obligatory drug-abusing behaviors as well as the creation of a physical dependence that might be best treated by restorative as well as pharmacological techniques.

Most of the early forms of medical treatment have long been thrown out, and also deemed uncommon and also cruel punishments. The overriding style today is that addiction can be dealt with through far more humane methods.

Several treatment versions may have their roots in previous techniques, nonetheless. Aversion therapy is still used today, although frequently through medicines like disulfiram that prevent people addicted to alcohol from alcohol consumption, as well as naloxone, an opioid agonist that can precipitate opioid withdrawal if mistreated, therefore assisting to avoid relapse.

Detox is typically achieved in either an outpatient or household setting, relying on the severity of an individual's dependence on drugs or alcohol. The more extreme the dependency, the more extreme withdrawal signs and medicine desires might be. A lot more extreme signs and also yearnings gain from medical detox, with continuous clinical monitoring, care, and the addition of pharmaceuticals.

Various other medication treatment programs beyond detoxification include:
- Prevention programs
- Outpatient programs
- Intensive outpatient programs
- Transitional programs
- Residential treatment programs
- Aftercare and recovery support programs

In hospitals, community centres, and organizations which provide the public with information on the dangers of drug abuse as well as dependency, preventative versions are often seen. We seek to help people seek care if possible. Outpatient, comprehensive ambulatory, transition and residential therapy programs, each with a different framework and identity, may consist of several the same research methods.

A domestic program most likely gives the highest level of care with 24-hour supervision and also a high degree of the framework.

Behavioral treatment sessions are consisted of in many drug treatment programs and normally include both group and individual sessions to introduce healthy coping mechanisms, and a much better understanding of the interaction in between thoughts and also actions and just how to customize them in favorable means. Abilities educating sessions as well as instructional opportunities might likewise be vital parts of modern drug therapy programs, as are counseling sessions.

Medication dependency treatment programs have come to a long method for many years and remain to advance as new research as well as scientific proof emerged. Today, typical and all-natural approaches are frequently blended to develop a comprehensive treatment model that seeks to help people, and also households strike an equilibrium between the mind, body, and soul while dealing with the entire individual. Nutrition and also physical wellness might be straight tied to psychological health. By dealing with both mental and physical health and wellness, a strong recuperation foundation can be created.

THE TRIUMVIRATE REASONS WHY TEENS SEEK DRUG ABUSE

Dating back in the mid-'60s or in the last stages of the 70's drug abuse has ended up being an irritating issue of society. It resembled opening up the well-known Pandora's Box, releasing every feasible wickedness in the whole world. The problem of substance abuse has survived the examination of time as well as the attempt of every human getting on eliminating it. People back then used medicines such as (weed) cannabis and also feasible opium or heroin. Their formulations are taken into consideration moderate compared to the medications kids make use of today. Medicine issue is a tried and tested hazard of society and in spite of all the efforts of authorities to control it, the majority of the time, their hard work and also perseverance would certainly be shed.

Battling your method back from medicine abuse is devastating as well as a lengthy process. Drugs have actually ruined millions of lives as well as households around the world. Some children acknowledge the use of medications as a means to be included in a certain team.

Reasons why children lean on medications: More on psychological

Many kids looking for help with illegal substances have only a problem in their households. Teenagers with mold and mildew are typically medically associated with a breakup family. The sooner the adolescent takes drugs, the greater his or her potential effects.

Effects of society

Social factors are taken into consideration to a basic aspect of medication addiction. Once again, family members would remain in the facility of the crossfire. Even at that time, peer pressure is constant unmovable pressures that either make you or break you. Essentially teens are attention applicants; if their moms and dads could not supply it regularly, they would certainly seek from an outside resource. It could be in the kind of pals, groups, as well as the awful part gangs.

Naturally, it is in you.

If medication dependency is a typical trouble in your family members, then it is considered to drop under biological variables. A background of medicine abuse in the family ought to be dealt with as soon as possible.

THE MIND AS WELL AS BODY RELATIONSHIP MENTAL ILLNESS WITH A METAPHYSICALLY EXPLANATION

In "mental disease," perpetuated by the psychiatric community, I want to explain the misuse of the term "mental." The term is openly used by practitioners in mental health and is misleading. The definition of mysticism is different from those of psychiatric disorders, particularly in my knowledge of metaphysical philosophy. I assume that all psychiatric patients have perfect mind and are regular. Given what medical professionals say to the contrary, I make this seemingly provocative statement. I entirely disagree with the view that circumstances over which a person has little power permanently damage his / her mind. Always the mind remains unchanged. Something "mental" about it. There's nothing. For people with actual mental illnesses, especially those involving depression, organic brain diseases are likely to be present. That's why psychiatric drugs can help manage the symptoms of these conditions often.

The functioning of the brain is compromised because of mental disorders. The subconscious knows the knowledge sent by the malfunctioning brain. It makes the mind of a human seem unlike other people. Their actions violates social rules that have been recognized. This is not the fault of the mind. Snafu is a biological problem. I would also like to note that intellect is not a reliable tool in measuring a person's health. Not to mention that in addition to traditional and scientific forms there are different kinds of intelligence.

It must be remembered that a person who is ill-equipped is unable to internalize him or her in a safe way because of traumatic experiences.

A trauma or a series of bad experiences that cause general anxiety, panic syndrome, and post-traumatic stress disorder.

These disorders are considered mental diseases under the DSM-V (the medical community's bible). The event(s) itself will lead the mind to unhealthily interpret reality. Psychiatric drugs are less effective and are best treated with speech therapy in people with these mental conditions. Remember that my opinions on this issue do not (if at all) fully agree with medical researchers ' findings. Your views on this subject are not fully acceptable to me. I see it from an analytical and philosophical viewpoint. The body has structures and chemicals that block our view of the often frightening things that we usually see. Our body is some form of sensory frustration. It is intended to isolate us from our roots so that we can focus solely on physical reality. The five senses, which are designed to facilitate our physical experience, provide all the stimulation we receive.

A chemical-induce portal to the spirit world is created by the introduction of psychoactive substances like LSD and Mescaline into the body. These psychedelic medications are often used for "tript" recreationally, allowing users to experience parts of the other side of what psychologists and other doctors rely on are hallucinations. They are regarded as having no substance or truth. Current psychics (and those in the past) are mindful of the energy of emotions, like men.

Psychotropic drugs suppress the so-called hallucinatory hallucinations in schizophrenia and other psychotic disorders. Drugs such as Haldol, Zyprixa, Risperdal, and Latuda can have an effect on people who have mental disorders as opposed to the reaction of psychoactive chemicals. To mentally ill patients, these medications are prescribed as a consequence of their disease. Such drugs inhibit the function of some brain chemicals as dopamine, which in some unknown way, gives the person their unusual experience. It can be distressing to see the subconscious representations outside our senses. These photos are weird and new to us. An unusual condition is triggered by the damaged brain, preventing a reasonable connection between the patient and others. They don't plant their sensory perception fully into physical reality.

The visual and audio experiences of a mentally ill person or recreational psychedelic drug user are very real, conscious, and intellectual. We are separate and distinct from our own feelings, even though they were our own at first. I am sure that the sensory disturbances that medical professionals speak about inadequately are not hallucinating. Influential writers of psychic literature throughout history developed the legitimacy of this supernatural theory.

CONCLUSION

Anthropological findings and personal records indicate that people have used psychedelic drugs for many thousands of years. It's a partial rundown of just a few of the oldest psychoactive substance users (Sullivan and Hagen, 2002).
- 13000 y.a.-- The betel nut was eaten in Timor.
- 10700 y.a.-- The betel nut was chewed in Thailand.
- Before European call-- The Aborigines utilized pituri and also Nicotiana (both pure nicotine).
- Before European call-- Native Americans made use of cigarettes.
- Before European get in touch with-- Ethiopians utilized khat.
- 7000 y.a.-- In the Andes, coca was used.
- 5000 y.a.-- Coca was being utilized in Equador

The extensive nature of psychoactive compound use by human beings is quite interesting, considering what we know now concerning the hazardous nature of such substances. The use of substances longs back is fairly various from the types of usages today. In societies where the above medications are used, the substances are not checked out as drugs; however, like food, which is easy to think, seeing as they do not come in modern-day purified forms, yet as plants (Sullivan and also Hagen, 2002).

Some users of psychedelic substances by ancient people are additionally extremely comparable to their uses today. They could be used as medications ("medicine usage") and also in spiritual events ("medication cult"). From there, Hippocrates, Assyrians, and Muslims used them as medicine, and many people used opium as a drug ("drug use").

Hallucinogens have a background that goes back practically 2000 years ("medicine usage"). The hallucinations produced by these substances often appear religious in nature, leading to their eventual usage in a spiritual context ("medicine cult").

It has actually been utilized for its medicinal characteristics for many of its usage, including today ("medicine use"). It was made use of by a people in the Altai Mountains as well as today it is made use of in religious techniques by individuals in India and also Africa ("drug cult").

Some professionals today also think that psychoactive medications are accountable for elements of some societies. Marlene Dobkin de Rios is of the view that psychoactively active substances are important to the faith and art of Maya (Dobkin de Rios et al. 1974). The religion of the Huichol Indians in Mexico is based on taking in peyote where they interact with their gods. It has additionally played a huge part in their art (Haviland). In these two cultures, a psychoactive compound was a major contributor to culture. With such a long background and extensive usage, who recognizes what kinds of influences these substances could have carried human worlds throughout the years?

Humans are also not the only psychedelic drugs used by animals. Circuits have taken part in a kind of drunkenness using these things (Siegel), like crows, Dogs, Elephants, and monks. These events of use by both people and also animals increase numerous inquiries concerning the nature of animal-drug connections. Especially, they raise questions about how plants and also pets evolved to ensure that certain plants have the effects that they do on the pet nervous system. They additionally make us wonder why an animal would be driven to utilize such substances. The video mentioned below provides an interesting comparison between the psychoactive and primate family activities.

Do Not Go Yet; One Last Thing To Do
I would be very happy if you would give a short review of Amazon if you liked the book or found it useful. Your encouragement really makes a difference, and I personally read all reviews to get your input and develop the book.

www.ingramcontent.com/pod-product-compliance
Lightning Source LLC
Chambersburg PA
CBHW050002230526
45465CB00003BB/1221